Sequential Competitive Location on Networks

AF764820

European University Studies

Europäische Hochschulschriften

Publications Universitaires Européennes

Series V	**Economics and Management**
Reihe V	Volks- und Betriebswirtschaft
Série V	Sciences économiques, gestion d'entreprise

Volume/Band **3427**

Dominik Kreß

Sequential Competitive Location on Networks

ACADEMIC
RESEARCH

Bibliographic Information published by the Deutsche Nationalbibliothek
The Deutsche Nationalbibliothek lists this publication in the Deutsche Nationalbibliografie; detailed bibliographic data is available in the internet at http://dnb.d-nb.de.

Zugl.: Siegen, Univ., Diss., 2013

D 467
ISSN 0531-7339
ISBN 978-3-631-62767-9

© Peter Lang GmbH
Internationaler Verlag der Wissenschaften
Frankfurt am Main 2013
All rights reserved.
PL Academic Research is an Imprint of Peter Lang GmbH.

All parts of this publication are protected by copyright. Any utilisation outside the strict limits of the copyright law, without the permission of the publisher, is forbidden and liable to prosecution. This applies in particular to reproductions, translations, microfilming, and storage and processing in electronic retrieval systems.

www.peterlang.de

To my grandfather Rudi

Preface

This dissertation would not have been possible without the support of numerous people, to all of whom I am deeply grateful.

First of all, I wish to express my gratitude to my advisor Prof. Dr. Erwin Pesch. His encouragement, guidance, support, as well as his constructive and detailed feedback have been the most valuable assets during the last years.

I would also like to thank Prof. Dr. Dirk Briskorn for serving as the second referee of my thesis and Prof. Dr. Dr. h.c. Wolfgang Domschke for awakening my interest in the field of Operations Research and paving my way to the research community.

I am very grateful to my colleagues and friends Dr. Sergei Chubanov, Roswitha Eifler, Prof. Dr. Florian Jaehn, Sebastian Meiswinkel, Dr. Jenny Nossack, Dr. Alena Otto, Daniel Schnitzler and Philipp Zeise at the University of Siegen, as well as Dr. Christian Lang, Dr. Daniel Scholz and Dr. Florian Seeanner and at the TU Darmstadt. You have been a fruitful source of encouragement and inspiration. Thank you for your valuable ideas and feedback.

I am thankful to my family, especially my parents, grandparents and the Helland and Huckins family for their support at all times and for bringing me up to think for myself. Also, I am deeply grateful to my friends Alex, Caspar, Johannes, Max and Timo for always being there. Thank you, Anika, for all your love and support as well as your patience (not only) during this project. Last but not least, I am thankful to the Eintracht Frankfurt family and the members of the EFC Hibbdebach for enriching my life with friendship and emotions.

Siegen, February 2013 Dominik Kreß

Contents

List of Figures

List of Tables

Chapter 1

Introduction and Preliminaries

Location problems are concerned with the location of (physical or nonphysical) resources in some given space. *Competitive location models* additionally incorporate the fact that location decisions have been or will be made by independent decision-makers who will subsequently compete with each other, e.g. for market share when we think of locating facilities such as gas stations or supermarkets (see Plastria, 2001; ReVelle and Eiselt, 2005, for similar definitions). The location space under consideration does not necessarily need to be of geographical nature: political parties, for example, are concerned with locating in issue spaces; products may be positioned in characteristics spaces.

The practical relevance of analyzing competitive location models arises from the fact that, when locating resources in competitive environments, enterprises typically make strategic decisions that play a crucial role for their long-term success or failure. In the context of an industrial enterprise, for example, the choice of locations for production facilities influences the long-term structure of the logistics network, as relocating will, in general, induce high cost. After making its locational decisions, the enterprise will have face up to the competition.

This thesis takes a mathematical perspective (or, more specifically, an operations research perspective) on location theory. Well known non-competitive frameworks in this field are, for example, concerned with the choice of optimal points in networks according to the minimization or maximization of the sum of weighted distances to the vertices (median or antimedian problems) or the minimization of the maximum or maximization of the minimum weighted distance to any vertex (center or anticenter problems). Classical competitive location problems include the $(r|p)$-centroid or the $(r|X_p)$-medianoid problem. This research analyzes extensions to and variations of these latter problems. It focuses on incorporating proportional choice rules, non-discrete demand, or additional pricing decisions of firms. Furthermore, it provides insights into the computational complexity of some of the resulting problems and proposes adequate solution methods.

1.1 Basic Notation and Definitions

In this thesis we will denote the set of natural numbers including zero by $\mathbb{N}$, the set of positive natural numbers by $\mathbb{N}^+$, the set of rational (real) numbers by $\mathbb{Q}$ ($\mathbb{R}$), the set of positive rational (real) numbers by $\mathbb{Q}^+$ ($\mathbb{R}^+$) and the set of nonnegative rational (real) numbers by $\mathbb{Q}_0^+$ ($\mathbb{R}_0^+$).

We assume the reader to be familiar with the fundamental concepts of operations research. The basic graph-theoretic definitions, along with the corresponding notation used throughout this thesis, however, are introduced in Section 1.1.1. Most of the definitions are taken from Alstrup et al. (2004); Bandelt (1985); Bauer et al. (1993); Garbe (1995); Gross and Yellen (2004); Swamy and Thulasiraman (1981). Similarly, Sections 1.1.2 and 1.1.3 are concerned with the basics of game theory and discrete choice theory. The latter Section is based on Train (2009).

1.1.1 Graphs and Networks

A *graph* $G = (V, E)$ is composed of two (finite) sets V ($n := |V|$) and E ($m := |E|$). The elements of V are called *vertices*. The elements of E are the *edges*. Each edge is associated (incident) to one or two vertices, which are called the edge's end points. The end points are said to be joined by the edge. An edge $e \in E$ joining vertex $u \in V$ and vertex $v \in V$ is denoted by $e = [u, v]$. Unless otherwise stated, we assume that there is no direction associated with an edge of a graph, hence $[u, v] = [v, u]$. Two vertices $u, v \in V$, $u \neq v$, are called *neighbored*, if there exists an edge $[u, v]$ in E. An edge $[u, u]$ is a *loop*. The sum of the number of edges that join a vertex $v \in V$ with other vertices of the network and twice the number of loops at v is the *degree* of v, $deg(v)$.

A *network* $N = (V, E, \lambda)$ consists of a graph with vertex set V and edge set E and an additional mapping $\lambda : E \rightarrow \mathbb{R}^+$. For each $e = [u, v] \in E$, $\lambda(e) = \lambda(uv) = \lambda(u, v)$ defines the *length* of e. We define $\hat{D} := \max_{e \in E} \lambda(e)$. The elements of the edges of a network N, including all the vertices, are the *points* x of N ($x \in N$). A *subedge* $[x, y]$ (or xy) of an edge $e \in E$ is determined by two points x and y on e ($x, y \in e$). The length of a subedge $[x, y]$ is denoted by $\lambda([x, y]) = \lambda(xy) = \lambda(x, y)$. A subedge defined by all points of an edge $[u, v] \in E$ without including the vertices u and v is denoted by (u, v).

Let $N = (V, E, \lambda)$ and $V' \subseteq V$. Then $N' = (V', E', \lambda')$ is the *subnetwork* of N on the vertex set V', if $E' \subseteq E$ consists of all those edges of N whose end points are in V'. The mapping λ' is the restriction of λ to E'.

A *path* $P(x, y)$ joining two points x, y, $x \neq y$, of a network N is either a subedge $[x, y]$ or an alternating sequence of (sub-) edges and vertices, $[x, v_0], v_0, e_1, v_1, ..., e_k, v_k, [v_k, y]$, such that $e_j = [v_{j-1}, v_j] \in E$ for $j = 1, ..., k$, no edge and no vertex of the sequence occurs more than once, and $[x, v_0]$ as well as $[v_k, y]$ define point-disjoint subedges of N. x and y are called end points of $P(x, y)$. The length of a path is defined by the sum of the lengths of its edges and subedges. A path $P(x, y)$ is a shortest path, if there exists no path $P'(x, y) \neq P(x, y)$ of smaller length; The length of a shortest path $P(x, y)$ is the *distance* $d(x, y) = d_{xy}$ between x and y. We define $d(x, y) := \infty$ if there exists no path $P(x, y)$ for $x, y \in N$, $x \neq y$. Furthermore, we define $d(x, x) := 0$ for any $x \in N$ and $D(p, Z) := \min\{d(p, z) | z \in Z\}$ for a point $p \in N$ and a set of points $Z \subseteq N$. A *cycle* consists of an edge $e = [u, v] \in E$, and some path $P(u, v) \neq e$ connecting u and v. In a *connected network* there exists a path joining x and y for any two points $x \neq y$ of the network. We additionally define any network with $n = 1$ to be connected. A connected network is a *tree network* if there exists no cycle or loop. A tree network where every vertex is end point of at most two edges is a *chain network*. A network without loops where every pair of vertices is joined by a unique edge is a *complete network*.

Let u and v be vertices of a network N. Then $I(u,v) := \{x \in N | d(u,v) = d(u,x)+d(x,v)\}$ is called the *interval* $I(u,v)$ between u and v.[1] The interval is *ported* if for any pair of points $x \in I(u,v)$ and $y \notin I(u,v)$ every shortest path $P(x,y)$ passes through u or v.

A *path-decomposition* of a graph $G = (V,E)$ is a sequence $V_1, ..., V_r$ of subsets of the vertex set V, such that

1. $\bigcup_{1 \le i \le r} V_i = V$,
2. there exists a V_i, $i \in \{1, ..., r\}$, such that $u \in V_i$ and $v \in V_i$ for all $[u,v] \in E$, and
3. $V_i \cap V_k \subseteq V_j$ holds for all $1 \le i < j < k \le r$.

The *width* of a path-decomposition $V_1, ..., V_r$ is defined as $\max_{1 \le i \le r}\{|V_i| - 1\}$. The *pathwidth* of a graph $G = (V,E)$ is the minimum width over all path-decompositions of G.

Let a tree network $N = (V, E, \lambda)$ be *rooted* at some distinguished vertex $r \in V$. For each pair of vertices $i \in V$ and $j \in V$, we call i a *descendant* of j, if j is on the unique path that connects i to the root r. If i is a descendant of j, we call j an *ancestor* of i. A vertex $v \in V$ is a *common ancestor* of two vertices $x, y \in V$, if it is an ancestor of both, x and y. A common ancestor of two vertices $x, y \in V$ is the *nearest common ancestor*, $nca(x,y)$, of these very vertices, if its distance to the root is the largest among all common ancestors. If $i \in V$ is a descendant of $j \in V$ and $[i,j] \in E$, then i is said to be a *child* of j and j is called the *father* of i. A vertex without children is a *leaf* of the tree network. For any vertex $v \in V$ we denote the subnetwork (subtree) of N on the vertex set $V_{T_v} := \{v\} \cup \{i \in V | i \text{ is a descendant of } v\}$ by T_v and the subnetwork on the vertex set $V'_{T_v} := V_{T_v} \setminus \{v\}$ by T'_v.

Unless otherwise stated, we assume that the networks considered in this thesis are connected and that there are no multiple edges. Moreover, we

1 Note that we denote open intervals of real space by (a,b), $a, b \in \mathbb{R}$, $a \le b$. Similarly, closed intervals of real space are denoted by $[a,b]$, $a, b \in \mathbb{R}$, $a \le b$.

assume that there are no loops at the vertices. We associate a (local) coordinate $x_{uv} \in [0, \lambda(uv)]$ with every edge $[u,v]$ of a network N. Thus, we are able to define any point of the network. The direction of counting can be defined arbitrarily. For a point p on edge $[u,v]$, we refer to the local coordinate's value at p by p as well, i.e. $x_{uv} = p$ at point p. It will always become clear from the context wether we refer to the point itself or to its corresponding coordinate's value. Finally, we assume the distances between all pairs of vertices of a network to be input data of our algorithms. These distances can, for instance, be computed in $\mathcal{O}(n^3)$ time with the Floyd-Warshall algorithm (see, for example, Aho, 2004).

1.1.2 Game Theory

Noncooperative game theory deals with the behavior of agents (or players) in situations where each agent's optimal choice depends on the (forecasted) choices of his opponents and where each agent is motivated solely by self-interest (Fudenberg and Tirole, 1991).

We will consider games in *strategic form* that have three basic elements (Fudenberg and Tirole, 1991): a set of players Θ which we assume to be finite ($\Theta = \{1, ..., \theta\}$), a (pure) strategy space Ψ_i for each player $i \in \Theta$, and a payoff function $u_i(\boldsymbol{\psi})$ for each player $i \in \Theta$ that assigns a utility level to every vector of strategies $\boldsymbol{\psi} = (\psi_1, ..., \psi_\theta)$, $\psi_i \in \Psi_i$.

A strategy vector $\boldsymbol{\psi}^N = (\psi_1^N, ..., \psi_\theta^N)$ is said to be a *Nash equilibrium* in pure strategies, if no player can unilaterally increase his utility, i.e. $u_i(\boldsymbol{\psi}^N) \geq u_i(\psi_i, \boldsymbol{\psi}^N_{\Theta\setminus\{i\}})$ for all $\psi_i \in \Psi_i$, where $\boldsymbol{\psi}^N_{\Theta\setminus\{i\}} = (\psi_j^N | j \in \Theta, j \neq i)$ (cf. Gabay and Moulin, 1980).

The following theorem is well known. We refer to Fudenberg and Tirole (1991) and the references therein for a proof.

Theorem 1.1.1. *Let Θ be a nonempty set of players and consider a strategic-form game whose strategy spaces Ψ_i, $i \in \Theta$, are nonempty, convex and compact subsets of an Euclidean space. If the payoff functions u_i are continuous in $\boldsymbol{\psi}$ and quasiconcave in ψ_i, then there exists a pure strategy Nash equilibrium.*

1.1.3 Discrete Choice Theory

Consider the choices of decision makers (for example customers) among a set of alternatives (for instance a set of facilities), the so called choice set. If the alternatives are mutually exclusive and the choice set is finite and exhaustive, we call this framework a *discrete choice model.* If one assumes the decision makers to be utility maximizing and derives a specific discrete choice model from this assumption, the model is said to be a *random utility model* (RUM). For a detailed overview of RUMs and their features we refer to Anderson et al. (1992a); Train (2009).

To derive a RUM, one takes the point of view of a researcher and assumes the utility u_{ij} of a decision maker $i \in I$ (I denotes the set of decision makers) from choosing alternative $j \in J$ (J refers to the choice set) to be composed of a deterministic (i.e. observable) component v_{ij} and a stochastic component ϵ_{ij}, the latter being related to unobservable, utility affecting factors:

$$u_{ij} = v_{ij} + \epsilon_{ij}. \tag{1.1}$$

As the decision makers are assumed to be utility maximizing, the probability pr_{ij} for a decision maker $i \in I$ to choose alternative $j \in J$ is

$$pr_{ij} = Prob(u_{ij} > u_{ik} \; \forall \, k \in J, k \neq j) \tag{1.2}$$

Now, the RUM is derived by additionally providing a specific distribution of the random components. One of the most prominent RUMs which is well established in the economics, marketing and operations research literature (see also Hensher et al., 2005), the *multinomial logit model*, assumes the stochastic components to be independently and identical extreme value distributed (Gumbel distributed) with density

$$f(\epsilon_{ij}) = \left(e^{-\epsilon_{ij}/\sigma} e^{-e^{-\epsilon_{ij}/\sigma}}\right) / \sigma \tag{1.3}$$

and the cumulative distribution

$$F(\epsilon_{ij}) = e^{-e^{-\epsilon_{ij}/\sigma}}. \tag{1.4}$$

The variance is $\sigma^2\pi^2/6$, where σ is a scaling parameter. As σ approaches zero, the decision makers' choices become deterministic. The mean is $\sigma\gamma$, where γ is Euler's constant. Then, as shown in Appendix A, one derives a closed form expression of the choice probabilities of the multinomial logit model,

$$pr_{ij} = \frac{e^{v_{ij}/\sigma}}{\sum\limits_{k \in J} e^{v_{ik}/\sigma}} \quad \forall i \in I, j \in J. \tag{1.5}$$

1.2 Foundations of Competitive and Voting Location

The study of competitive location is rooted in the work of Hotelling (1929), who studied the location choice and pricing decision of two competitors on a finite line with uniformly spread consumers. It has been a prosperous field of research, leading to a whole variety of models, ever since. Those models can be classified with respect to multiple components (Eiselt, 1993; Eiselt and Laporte, 1989, 1996; Eiselt et al., 1993; Friesz et al., 1988; Hamacher and Nickel, 1998; Plastria, 2001; ReVelle and Eiselt, 2005). In this section we describe the most significant of these classification criteria as in Kress and Pesch (2012d).

Most important, the *representation of the underlying location space* traditionally gives rise to three classes: d-dimensional real space, network and discrete space. Distances need to be calculated by some metric in each of these classes. We follow ReVelle and Eiselt (2005) in differentiating only between d-dimensional real space and network location problems, each of which further being subdivided into continuous and discrete problems (see Figure 1.1). A *discrete* problem arises, when the set of candidate locations is assumed to be finite and known a priori. In a *continuous* problem, any point of the network or the d-dimensional space is a potential location site. By identifying finite dominating sets – finite sets of points to which at least one of the optimal solutions must belong – we are able to transform some special classes of continuous location problems into equivalent discrete problem classes a posteriori (Hooker et al., 1991). Moreover, as discrete sets of potential facility sites may easily become very large, one may consider treating those sets as continuous entities (see, for instance, Dasci and Laporte, 2005). Note, however, that there is an overlap of d-dimensional real spaces and networks. Hotelling's line, for instance, can be interpreted as a network with two vertices and a connecting edge or, alternatively, as an interval of $\mathbb{R}^1$. Other models are somewhere in between network and d-dimensional models. Suárez-Vega et al. (2011), for example, consider a "buffer" around a network to represent the space of potential locations.

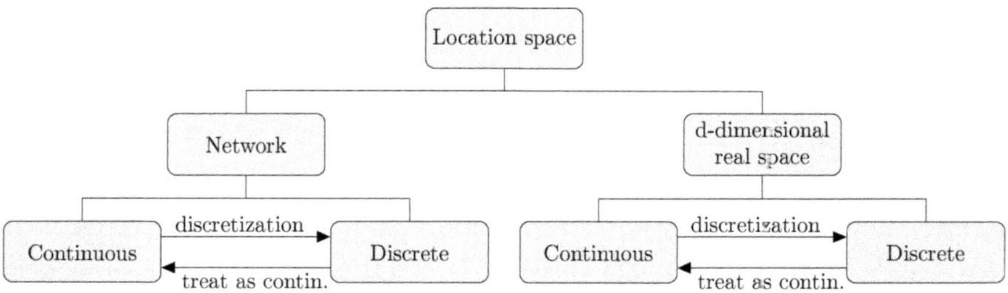

Fig. 1.1: Location spaces

Other fundamental categories of competitive location theory are related to game theoretic aspects. Competition itself may be static (present and fixed), competitors may enter in a simultaneous or sequential fashion, or we can think of dynamic competition, i.e. players who repeatedly reoptimize their locations (see Figure 1.2; Eiselt et al., 1993; Plastria, 2001). *Sequential locational competition*, dating back to Hay (1976) and Prescott and Visscher (1977), is characterized by two types of players: leaders, who choose locations at given instants, anticipating the subsequent actions of later entrants, and followers, who make their location decisions based on the past decisions of the leaders. The solution concept generally employed in sequential location problems is the Stackelberg equilibrium (von Stackelberg, 1934): assuming rational players, the location of each player is determined by backward induction. *Simultaneous locational games* (as the one of Hotelling, 1929), in contrast, usually use the concept of a Nash equilibrium (cf. Section 1.1.2). Note that the *number of players* may be exogenously given or determined endogenously, e.g. by incorporating fixed location cost. The same holds for the *sequence* of location and the *numbers of resources* to be located by each player.

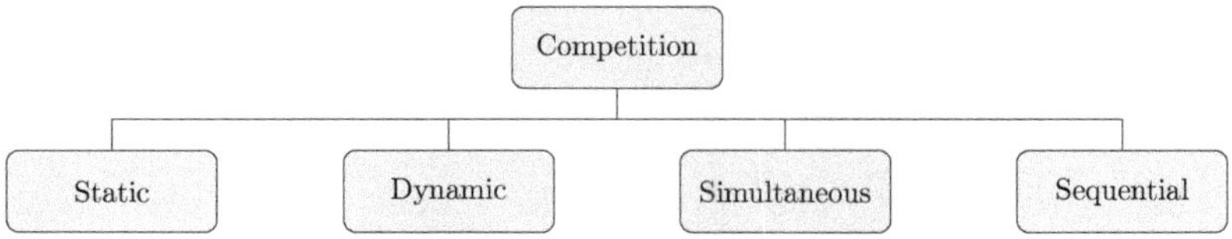

Fig. 1.2: Competition

It is generally agreed that the work on competitive location problems on (general) networks is rooted in the work of Slater (1975) and Hakimi (1983) (see Eiselt and Laporte, 1996; Smith et al., 2009, for details). Hakimi (1983) formally introduced the terms $(r|X_p)$-medianoid problem and $(r|p)$-centroid problem for sequential games with one leader (L) and one follower (F) locating p and r facilities, respectively. Note that r and p are arbitrary input parameters. Knowing the p locations of L, denoted by $X_p = (x_1, ..., x_p)$, F faces the problem of optimally locating r facilities (with respect to some objective function): the *$(r|X_p)$-medianoid problem.* We denote a feasible location decision of F by $Y_r = (x_{p+1}, ..., x_{p+r})$ and an optimal location decision by $Y_r^* = (x_{p+1}^*, ..., x_{p+r}^*)$. L's problem, the *$(r|p)$-centroid problem*, is to locate p facilities, anticipating F's subsequent behavior. An optimal solution to this latter problem is denoted by $X_p^* = (x_1^*, ..., x_p^*)$. Note that, differing from other authors as Spoerhase and Wirth (2010), we use the terms $(r|X_p)$-medianoid and $(r|p)$-centroid problem in a rather broad sense, subsuming a whole variety of choice rules and player objectives under these terms.

Another category that is related to game theory is the incorporation of pricing in competitive facility or product location models (Anderson et al., 1992b; Eiselt and Laporte, 1996; Eiselt et al., 1993; García Pérez et al., 2004). Traditional spatial pricing policies include *mill pricing* (all customers are charged the same price for the good itself, all transport costs are passed to the customers), *uniform delivered pricing* (the facility pays for the transportation; all customers of a facility are charged the same price, no matter where they are located at) and *spatial price discrimination* (the facility pays for the transportation; prices for different customers are customer-location-specific). Prices may be set simultaneously to the location decisions or in a separate stage, either sequentially or simultaneously. Alternatively, one can incorporate *parametric prices.* The equilibrium concepts used for combined location price games depend on which of these situations is implemented (Eiselt et al., 1993).

Other ingredients of competitive location models include *characteristics of the targeted group*, as, for example, customers or voters. They may be distributed over the representation of the location space according to some density function or we may consider discrete locations (cf. Section 1.3). Demand may be *deterministic* or *stochastic*, *elastic* or *inelastic* (dependent on or independent of the conditions of its supply). Furthermore, we may take different types of choice rules into consideration (cf., for example, Hakimi, 1990). A choice rule is said to be *binary* (or *deterministic*), when it is deterministic from the perspective of the players with the total demand of a customer (voter etc.) being served by a single located resource; it is said to be *probabilistic*, if the researcher can only derive probabilities of customer (voter etc.) behavior or if demand is assumed to be split over multiple located resources. Probabilistic choice rules include *partially binary* (splitting only over one of the locations of each player) and *proportional* (splitting over all locations) behavior. Note that, at least in the case of a binary choice rule, one has to make assumptions concerning the location of two facilities in the same point of the network (*co-location*). A common assumption in the field of competitive location problems is to break ties in favor of the leader (see, for example, Hakimi, 1990; Hansen and Labbé, 1988; Hansen and Thisse, 1981) or, similarly, not allowing co-location at all (see, for example, Granot et al., 2010; Shiode and Drezner, 2003). Alternatively, ties may be broken equally as, for instance, in Dasci et al. (2002). Both, the existence and nature of equilibria in competitive location models, may vary according to different assumptions concerning co-location: Hakimi (1990) designs the above-mentioned tie breaking rule to "avoid [...] trivial solutions". Similarly, Granot et al. (2010) analyze the effect of allowing or not allowing co-location in their model in detail.

A whole variety of other features may be incorporated into competitive location problems. These features include incomplete information: for example, information about marginal costs of production may be asymmetric. Moreover, different types of players pursue different types of objectives: public firms, for instance, maximize domestic welfare, while private firms maximize profits. We may consider multiple products or multiple markets as well.

Voting location problems, i.e. problems of locating resources as the result of a collective election, are closely related to competitive location problems (Hansen et al., 1990). Think, for example, of the problem of locating a school in a city and suppose that the benefit enjoyed by an individual decreases when the traveling distance to the facility becomes larger (as, for instance, assumed by Bauer et al., 1993). The council of the city of Birmingham in the United Kingdom, for instance, wishes to find "placements that best meet the needs of children and young people and the aspirations of parents and carers, as close as possible to where they live" (Birmingham Council, 2011). We are faced with the problem of finding a compromise within the group of residents. Even though the users vote in order to locate resources that can be interpreted to be owned by a single decision maker, the voting process is such that a stable location is generally characterized by the nonexistence of a strong party of users who prefer alternative locations, that may be seen as possible locations of a competitive decision maker. Thus, although the focus of voting location problems is somewhat different, they may be interpreted as a special case of sequential competitive location problems. There is a strategic element that these problems share (Hansen et al., 1990): independent decision makers (leaders and followers or voters) influence the solution by making their decisions.

Another class of closely related location problems is concerned with *coverage* objectives. These types of objectives model situations in which the users seek to have a resource within a reasonable distance. "There is [...] some competitive flavor to such problems in that [...] existing facilities may belong to one company while a second company is trying to extract the maximum profit by locating its own facilities" (Megiddo et al., 1983).

We will use the terms user, customer and voter, facility and resource, as well as the terms number of voters and demand interchangeably throughout this thesis. Furthermore, unless otherwise stated, we will consider location problems on networks and we will assume that the players have the objective of maximizing profit, which, in the absence of prices, becomes equivalent to maximizing market share. We will occasionally precede the terms $(r|p)$-centroid and $(r|X_p)$-medianoid problem with the terms discrete, continuous, binary or proportional to clarify the type of choice rule (binary or proportional) and location space (discrete or continuous problems on networks) under consideration.

1.3 Vertex and Edge Demand

Whereas competitive location models in $\mathbb{R}^1$ typically assume the demand to be continuously dispersed over the line segment, the majority of network models incorporate discrete demand, i.e. demand arising in the vertices of the network (see Chapter 2). Dasci et al. (2002) and Okunuki and Okabe (2002) were among the first to consider demand densities over the edges of competitive location problems on general networks. In an urban context, the motivation is as follows: Cities are typically modeled as networks. Edges correspond to streets and vertices represent intersections. The houses of a city are usually dispersed over the streets. Thus, demand is typically non-discrete.

In the remainder of this thesis, we will consider network models with vertex demand only, edge demand only or both, vertex- and edge demand. Unless otherwise stated, our notation is an extension of the notation used by Bauer et al. (1993) as presented in Kress and Pesch (2012c) (see Figure 1.3): Given a network $N = (V, E, \lambda)$, there is a finite number of users located at the vertices (vertex demand, vertex customers). There may be several users or no users at all at each vertex. Their demand is represented by a weight function $\pi : V \to \mathbb{R}_0^+$. We define $\pi(x) := 0$ for all $x \notin V$. Additionally, we consider a mapping $\delta : E \to \mathbb{R}_0^+$. The value $\delta(e) = \delta(uv) = \delta(u, v)$ is the uniform demand density (edge demand, edge customers) of the edge

$e = [u, v] \in E$, i.e. the demand per unit of length. δ and π may not be equal to the zero function at the same time.

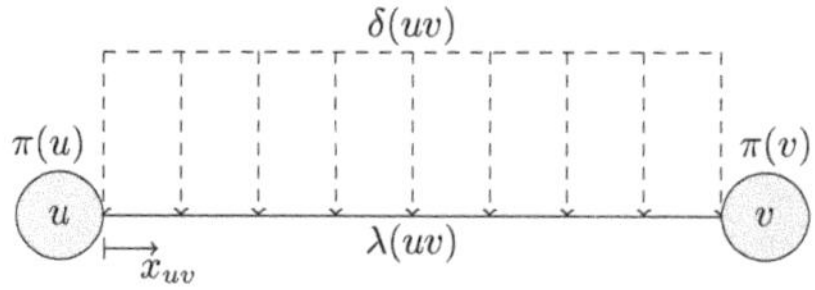

Fig. 1.3: Vertex and edge demand

For a subnetwork $N' = (V', E', \lambda')$ of N we denote by $\pi(V')$ or $\pi(N')$ the sum $\sum_{u \in V'} \pi(u)$. Analogously, we denote by $\delta(E')$ or $\delta(N')$ the sum $\sum_{e \in E'} \delta(e) \cdot \lambda(e)$. Finally, we define $\xi(N') := \pi(V') + \delta(E')$.

Consider an $(r|X_p)$-medianoid or (r, p)-centroid framework. Unless otherwise stated, we denote the accumulated demand (or expected demand) of the (vertex and edge) customers who accommodate their demand at the follower's facilities by $W_F(Y_r(X_p)|X_p)$, where $Y_r(X_p)$ corresponds to an arbitrary feasible set of the follower's locations, given the location decision X_p of the leader. The optimal follower's market share is denoted by $W_F^*(Y_r(X_p)|X_p)$. Analogously, we denote the leader's (optimal) market share by $W_L(r|p)$ ($W_L^*(r|p)$).

1.4 Outline of the Thesis

The remainder of this thesis is organized as follows. In Chapter 2 we present a review of recent contributions in the field of sequential competitive location problems. It extends a well known review paper by Eiselt and Laporte from 1996. Chapter 3 is concerned with a competitive location problem on networks under a proportional choice rule that has previously been introduced by Bauer et al. (1993). We provide new insights into the characteristics of related point sets. Furthermore, we are concerned with implementational issues of an algorithm proposed in the original publication and we extend the original model to include edge demand. Chapter 4 is devoted to the binary $(r|p)$-centroid problem on networks with vertex and edge demand. We analyze its time complexity and basic properties. Moreover, we present an algorithm for the discrete problem class on tree networks with one follower facility only. In Chapter 5 we analyze the effect of including price competition into the discrete $(r|X_p)$-medianoid problem on networks, where proportional customer behavior is modeled by applying the multinomial logit approach. The chapter is concerned with time complexity issues and the existence of price equilibria. We analyze standard numerical approaches to determine local price equilibria and provide first insights into the sensitivity of the optimal locations of facilities to the level of price competition. Simple heuristic algorithms for finding solutions to the location problem are proposed. The thesis closes with a summary and an outlook on future research in Chapter 6.

All computational tests in this thesis are run on a Laptop with an Intel Core i7 CPU, 2.67 GHz, 4GB system memory, running under the 64bit Windows 7 Professional operating system.

Chapter 2

Sequential Competitive Location on Networks: Literature Review

Eiselt and Laporte presented a survey of sequential competitive location problems in 1996. Various publications in this field have appeared since then. We will therefore extend Eiselt and Laporte (1996) by reviewing some of these recent developments in this chapter as in Kress and Pesch (2012d). The scope of review is as follows:

- We restrict ourselves to detailed descriptions of publications that have appeared after Eiselt and Laporte (1996).
- We include voting location problems.
- Our focus is on models where potential locations are not a priori restricted to a finite set of points, as one finds up-to-date reviews concerning discrete models in Current et al. (2002); Daskin (2008); ReVelle et al. (2008); Serra and ReVelle (1995).
- Furthermore, we restrict our attention to problems defined on networks. Since a line, i.e. an interval of one-dimensional real space, may be interpreted as a special type of network and because models defined on lines might contain ideas worth adapting in more general network models, we include these models as well, yet without describing them in detail for the sake of brevity.

- We explicitly exclude location decisions in political issue spaces and refer the reader to Osborne (1995) and the references therein.
- Furthermore, for the most part, we exclude coverage problems as an excellent, very recent survey is Berman et al. (2010). Only indispensable models are surveyed in this chapter.

Tables 2.1 (competitive location problems) and 2.2 (other fields of location theory) give an overview of several selected review and classification papers that complement this review. We refer the reader to the books by Anderson et al. (1992a), Daskin (1995), Drezner (1995b), Drezner and Hamacher (2002), Eiselt and Marianov (2011), Handler and Mirchandani (1979), Love et al. (1988), Miller et al. (1996) and Mirchandani and Francis (1990) as well.

Tab. 2.1: Competitive location problems - reviews and classifications

Authors and year	Class	Type
Dasci (2011)	competitive location	classical contributions, review
Plastria (2001)	static competitive location	taxonomy and review
Eiselt and Laporte (1996)	sequential competitive location	taxonomy and review
Drezner (1995a)	competitive location in the plane	review
Serra and ReVelle (1995)	competitive location in discrete space	review
Eiselt et al. (1993)	competitive location	taxonomy and bibliography
Eiselt (1993)	competitive location	taxonomy
Eiselt and Laporte (1989)	competitive location	taxonomy and review
Friesz et al. (1988)	competitive location in networks	taxonomy and review
Graitson (1982)	competitive location on a line	review

This chapter proceeds as follows. Sections 2.1 and 2.2 are concerned with voting location problems. Sections 2.3 and 2.4 discuss selected $(r|X_p)$-medianoid and $(r|p)$-centroid problems, respectively. Section 2.5 is devoted to sequential competitive location problems that can not be interpreted in the sense of Hakimi's notion as described in Chapter 1.2. Furthermore, selected

Tab. 2.2: Other selected reviews and classifications

Authors and year	Class	Type
	Location problems in general	
Smith et al. (2009)	location	historical development
ReVelle and Eiselt (2005)	location	review
Hale and Moberg (2003)	location	review
Avella et al. (1998)	location	state of the art and future trends
Hamacher and Nickel (1998)	location	classification scheme
Domschke and Krispin (1997)	location	review
Brandeau and Chiu (1989)	location	overview
Domschke and Drexl (1985)	location	bibliography
Tansel et al. (1983a,b)	location	review
Francis and Goldstein (1974)	location	bibliography
	Discrete location problems	
ReVelle et al. (2008)	discrete location in special branches	taxonomy and bibliography
Daskin (2008)	discrete location	taxonomy and illustrations
Current et al. (2002)	discrete location	review
McGinnis (1977)	discrete location	review
	Continuous, d-dimensional location problems	
Plastria (2002)	continuous covering location	review
Plastria (1995)	continuous location problems	review
	Location problems with special application or context	
Melo et al. (2009)	facility location and supply chain management	review
Nagy and Salhi (2007)	location and vehicle routing	review
Şahin and Süral (2007)	hierarchical facility location	review
Snyder (2006)	stochastic and robust facility location	review
Klose and Drexl (2005)	facility location for distribution system design	review
Owen and Daskin (1998)	strategic facility location	review
Osborne (1995)	spatial political competition	review
Mesa and Boffey (1996)	extensive facility location in networks	review
Erkut and Neuman (1989)	undesirable facility location	classification scheme and review
Aikens (1985)	facility location for distribution planning	review

models on a line are described. The sensitivity of sequential competitive location problems with respect to the underlying modeling assumptions is subject of Section 2.6. The chapter ends with a conclusion in Section 2.7.

2.1 Voting Location: Condorcet Points and Related Concepts

Consider a network $N = (V, E, \lambda)$ with (deterministic and inelastic) vertex demand and let $x, y \in N$ be any pair of points (locations) of the network. A customer is indifferent about the locations x and y, denoted by $x \sim y$, if the difference of their distances to the customer is within a given threshold $\alpha \geq 0$ (equal for all customers in the network). Otherwise, the customer is said to prefer ($\succ$) one of the locations. If he prefers x to y, we write $x \succ y$. Now partition the vertex set V into three sets as follows:

$$[x \succ y] := \{u \in V | d(u, x) + \alpha < d(u, y)\}, \tag{2.1}$$

$$[x \sim y] := \{u \in V || d(u, x) - d(u, y)| \leq \alpha)\}, \tag{2.2}$$

$$[y \succ x] := \{u \in V | d(u, x) - \alpha > d(u, y)\}. \tag{2.3}$$

Let $\alpha = 0$ and define another parameter $0 \leq \beta \leq 1$. Then a point $x \in N$ is a *β-Condorcet point* (Bandelt, 1985), if

$$\pi([y \succ x]) + \beta\pi([x \sim y]) \leq 0.5\pi(N) \qquad \forall y \in N. \tag{2.4}$$

Hence, a β-Condorcet point is any point x with no strict majority of customers preferring another point to x, where β defines the behavior of indifferent customers. β-Condorcet points are called *Condorcet points* (Hansen and Thisse, 1981) for $\beta = 0$, *plurality points* (Wendell and McKelvey, 1981) for $\beta = 0.5$ and *majority points* (Wendell and Thorson, 1974) for $\beta = 1$.

It is well known that both, the structure of the network as well as the distribution of the users, influence the existence of β-Condorcet points (see, for instance, Bandelt, 1985; Hansen and Thisse, 1981; Hansen et al., 1986; Labbé, 1985). In tree networks, for example, the set of Condorcet points and medians is equal, so that there always exists at least one Condorcet point. This, however, needs not be true in general networks. Bandelt (1985) derives conditions (for all user distributions) under which general networks always have a Condorcet point and under which Condorcet points and medians coincide. Hansen and Labbé (1988), among others, present an algorithm of polynomial time complexity to determine the set of Condorcet points of a given network.

The Condorcet concept has been taken some steps further during the last decade. Campos Rodríguez and Moreno Pérez (2000a) define *α-Condorcet points* by allowing threshold values $\alpha \geq 0$ (and thus using an extended preference structure compared to the one used for the definition of Condorcet points) and setting $\beta = 0$ in (2.4). Obviously, there always exists a value for α, so that at least one α-Condorcet point exists. The authors give tight bounds for the relation of the objective function of median and center problems at an α-Condorcet point and the optimal values of the objective functions at a median or center in general networks and trees. These results extend similar well known results given in Hansen and Thisse (1981) and Labbé (1985) for Condorcet points. Campos Rodríguez and Moreno Pérez (2000b) modify the algorithm presented by Hansen and Labbé (1988) to obtain a polynomial algorithm to determine the set of α-Condorcet points for a given α.

The minimum α value that guarantees the existence of an α-Condorcet point is called tolerance distance. The corresponding α-Condorcet point, the *tolerant Condorcet point*, is introduced by Campos Rodríguez and Moreno Pérez (2000b). The authors provide a polynomial time algorithm to compute the tolerance distance of a given network. One of their main theoretical results states, that every point at a distance less than α from a Condorcet point is an α-Condorcet point. This, however, is not a necessary condition for α-Condorcet points.

Campos Rodríguez and Moreno Pérez (2003) extend the definition of α-Condorcet points to *$\alpha\gamma$-Condorcet points* by using

$$\pi([y \succ x]) + \beta\pi([x \sim y]) \leq \gamma\pi(N) \qquad \forall y \in N \tag{2.5}$$

instead of (2.4), where $0 \leq \gamma \leq 1$.[1] Thus, the proportion of customers needed to reject a location may be different from one half. If we fix only γ and compute the tolerance distance for this case we get *γ-tolerant Condorcet points*. Analogously, one may define *γ-Condorcet points* by setting $\alpha = 0$ or other point sets as, for example, *$\alpha\gamma$-plurality points* (setting $\beta = 0.5$) (Campos Rodríguez and Moreno Pérez, 2008). Unfortunately, the algorithms and the majority of the results presented in Campos Rodríguez and Moreno Pérez (2003) and Campos Rodríguez and Moreno Pérez (2008) are restricted to the case, where the number of possible facility locations is a priori assumed to be finite, and thus lie beyond the scope of this review. Nevertheless, note that the point sets defined in this section are extended to include the case of multiple facilities in Campos Rodríguez and Moreno Pérez (2008). For example, a *p-Condorcet* set is a set of p locations such that there is no other set of p locations that is preferred by a strict majority of the customers.

Another contribution dealing with discrete versions of (single and multiple facility) $\alpha\gamma$-Condorcet and related problems is due to Noltemeier et al. (2007). The authors suggest an extension of (2.1)–(2.3) by defining more general *threshold functions* $\delta : \mathbb{R}_0^+ \times \mathbb{R}_0^+ \rightarrow \mathbb{R}_0^+$, where $\delta(0, y) = 0$ for all $y \in \mathbb{R}_0^+$ and δ is (weakly) monotonously increasing in both parameters. Given a threshold function δ and a parameter $\beta' \geq 0$, they replace (2.3) (and analogously (2.1) and (2.2)) by

$$[y \succ x] := \{u \in V | d(u, x) - \delta(\beta', d(u, x)) > d(u, y)\} \tag{2.6}$$

for the single facility case and give a straight forward extension of $\alpha\gamma$-Condorcet points to *$\beta'\gamma$-Condorcet points*.

So far, we have only been concerned with deterministic choice rules. Bauer et al. (1993) define a proportional choice rule and extend the concept of Condorcet points to what they call *k-optimal points*, where $k \in \mathbb{N} \cup \{\infty\}$ is a parameter of the choice rule. For very large k the decisions of the cus-

1 Note that the term $\beta\pi([x \sim y])$ is not explicitly used by Campos Rodríguez and Moreno Pérez (2003), since $\beta = 0$ as it has been defined for α-Condorcet points. We include the term at this point to indicate that a β-value different from zero may still be considered, for instance by defining *$\alpha\beta\gamma$-Condorcet points*.

tomers are alike to the deterministic decision in the Condorcet case. We will describe their model and main results in detail in Chapter 3.

Figure 2.1 summarizes some of the relations of the point sets that have been mentioned in this section so far.

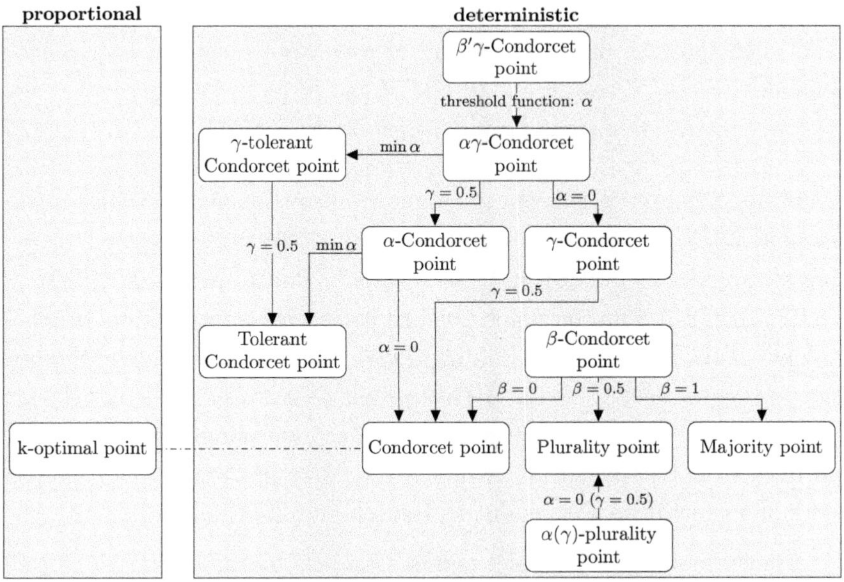

Fig. 2.1: Condorcet points and related concepts

We close this section by noting that Labbé (1990) defines *anti-Condorcet points* for the case where the players want to locate the facilities as far as possible from the users in the network. In this case no other point is allowed to be farther away from a strict majority of the users. The author presents a linear time algorithm for finding an anti-Condorcet point in trees. Furthermore, she gives a tight bound for the relation of the objective function of the antimedian problem at an anti-Condorcet point and at an antimedian. Finally, the *α-anti-Condorcet point* is defined analogously to the α-Condorcet point.

2.2 Voting Location: Some (1|1)-Centroid Problems

Consider the preference structure (2.1)–(2.3) with $\alpha = 0$. As stated in the previous section, β-Condorcet points do not always exist in general networks. This leads to the incorporation of a minimax objective, i.e. the search for points such that the maximal number of users who prefer another point is minimal:

$$\min_{x \in N} \{\max_{y \in N} \{\pi([y \succ x]) + \beta\pi([x \sim y])\}\}. \tag{2.7}$$

An optimal solution to this optimization problem is called *β-Simpson point* (Bandelt, 1985). β-Simpson points are called *Simpson points* if $\beta = 0$ (Hansen and Labbé, 1988) and are originally due to Simpson (1969). If $\beta = 0.5$, then a β-Simpson point is called a *security point* (Slater, 1975). In the view of Hakimi's notion (Hakimi, 1983), the different kinds of Simpson points are equivalent to those (1|1)-centroids that consider the corresponding choice rules. A general network contains at least one Simpson point. An algorithm of polynomial time complexity to determine the set of Simpson points of a general network is due to Hansen and Labbé (1988).

As in the case of α-Condorcet points, Campos Rodríguez and Moreno Pérez (2003) incorporate threshold values $\alpha \geq 0$ and set $\beta = 0$ to define *α-Simpson points.* Consider the definition of an $\alpha\gamma$-Condorcet point. An α-Simpson point is then defined to be an $\alpha\gamma^*(\alpha)$-Condorcet point, where $\gamma^*(\alpha)$ is the smallest possible value that γ may take, such that the set of $\alpha\gamma$-Condorcet points is not empty.

The definition of other point sets, as, for example, *α-security points* is straight forward (see Campos Rodríguez and Moreno Pérez, 2008). Again, note that Campos Rodríguez and Moreno Pérez (2008) and Noltemeier et al. (2007) consider the case of multiple facilities as well. However, the results presented in Campos Rodríguez and Moreno Pérez (2003, 2008) and Noltemeier et al. (2007) are mainly concerned with discrete location problems and are thus omitted in this chapter.

As, in the proportional model of Bauer et al. (1993), k-optimal points need not exist in general networks, the authors introduce *k-suboptimal points* in analogy to the Simpson concept. Details are presented in Chapter 3.

Figure 2.2 extends Figure 2.1 to give an overview of the point sets defined in this section so far.

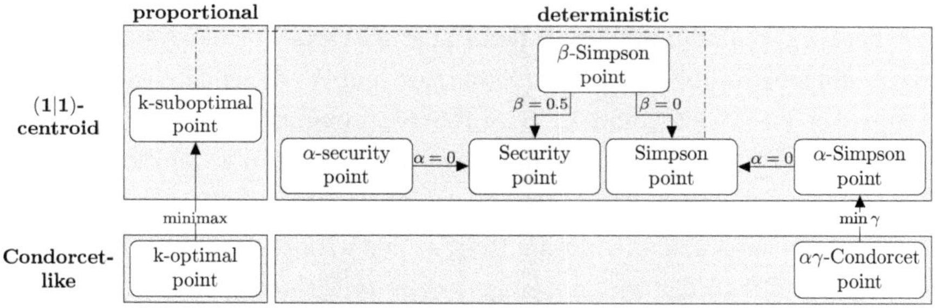

Fig. 2.2: Some (1|1)*-centroid problems related to the Condorcet concept*

Spoerhase and Wirth (2010) consider the preference structure (2.1)–(2.3) and introduce the concept of *monotonous gain functions* as a general model to describe several competitive and voting location problems. They consider tree networks $T = (V, E, \lambda)$ with $\pi : V \to \mathbb{Q}_0^+$, $\lambda : E \to \mathbb{Q}^+$ and $\lambda([u, v]) = d(u, x) + d(x, v)$ for any point x on an edge $[u, v] \in E$. A gain function $\Phi : T \times T \to \mathbb{Q}$ maps a pair of points (y, x) to the value $\Phi(y, x)$, denoted by $\Phi(y \succ x)$ as well, as a measure of the follower's influence at point y after the leader has located in point x. A gain function is called monotonous, if there is a function $\varphi : \mathbb{Q} \times \mathbb{Q} \to \mathbb{Q}$ such that

1. $\Phi(y \succ x) = \varphi(\pi([y \succ x]), \pi([x \succ y]))$ for all points $(x, y) \in T$,
2. φ is monotonously increasing in the first parameter and monotonously decreasing in the second parameter,
3. φ can be evaluated in constant time.

Moreover, the authors define the *absolute* Φ*-score of a point* x to be $\Phi(x) := \max_{y \in T} \Phi(y \succ x)$ and the *absolute* Φ*-score of a tree* to be $\Phi^* := \min_{x \in T} \Phi(x)$. Any point $y \in T$ where $\Phi(y \succ x) = \Phi(x)$ is called a Φ*-witness* of the leader point x. An *absolute Φ-solution* is a point $x \in T$ with $\Phi(x) = \Phi^*$. Therefore, the α-Simpson concept, for example, can be modeled by the monotonous gain function $\Gamma(y \succ x) := \pi([y \succ x])$. Thus, any absolute Γ-solution is an α-Simpson point. Spoerhase and Wirth (2010) derive an algorithm that finds an arbitrary element of the set of all absolute Φ-solutions for any monotonous gain function on a tree in time $\mathcal{O}(n)$. The algorithm iteratively decreases the tree network, maintaining a so called leader tree which is always guaranteed to contain an absolute Φ-solution. Spoerhase and Wirth (2009b) extend these results (for $\pi : V \to \mathbb{R}_0^+$ and $\lambda : E \to \mathbb{R}^+$) by providing an algorithm of time complexity $\mathcal{O}(n \log n)$ to compute the set of all absolute Φ-solutions for any monotonous gain function on a tree.

We close this section by noting that, in line with the anti-Condorcet concept, Labbé (1990) defines *anti-Simpson points* without giving any concrete results.

2.3 (r|X$_p$)-Medianoid Problems

Consider the characteristics of a *basic $(r|X_p)$-medianoid problem* where the leader's facilities are located at distinct vertices: Customers are solely located at the vertices of the network and select the closest facility (binary choice) to accommodate their deterministic, inelastic demand. Well known results on the computational complexity of this problem are listed in Table 2.3 (see, for example, Eiselt and Laporte, 1996, for more details). Megiddo et al. (1983) show that an optimal solution to this basic $(r|X_p)$-medianoid problem can be determined in $O(n^r m^r / r!)$ time by applying a simple enumeration procedure. This is based on the fact that there always exists an optimal solution at a finite set of so called boundary points (cardinality $\mathcal{O}(nm)$) of the network.

A similar discretization result for this basic $(r|X_p)$-medianoid problem has recently been presented by Suárez-Vega et al. (2004a) (cf. also Pelegrín et al., 2006, 2010; Santos-Peñate et al., 2007). The authors employ the concept of *isodistant points* in networks. This concept has been introduced by Peeters and Plastria (1998) (see below). A point x is said to be (u_j, f)-isodistant relative to a customer located at $u_j \in V$ with $\pi(u_j) \neq 0$ and an

Tab. 2.3: Some complexity results on a basic $(r|X_p)$-medianoid problem

Problem characteristics	NP-hard	Polynomial
customers located at vertices binary choice (closest facility) deterministic, inelastic demand	$(r\|X_p)$ on general networks (Megiddo et al., 1983) $(r\|X_1)$ on general networks (Hakimi, 1983)	$(r\|X_p)$: $\mathcal{O}(rn^2)$ on tree networks (Megiddo et al., 1983)

existing facility $f \in X_p$, if $d(u_j, x) = d(u_j, f)$. Any (u_j, f)-isodistant point for some $u_j \in V$ and $f \in X_p$ is called isodistant point. Pelegrín et al. (2010) present an algorithm of time complexity $\mathcal{O}(n^3)$ to determine all isodistant points of a network for known values of $d(u_j, f)$ for all $u_j \in V$ and $f \in X_p$. For the basic $(r|X_p)$-medianoid problem under consideration, Suárez-Vega et al. (2004a) prove that the follower's market share is constant on each of the open segments of the network, that are defined by neighboring vertices or isodistant points. Therefore, one can discretize the $(r|X_p)$-medianoid problem by considering an arbitrary point on each of these segments as well as the isodistant points and vertices of the network.

Suárez-Vega et al. (2004a) consider extensions of the basic $(r|X_p)$-medianoid problem as well (cf. also Santos-Peñate et al., 2007). For example, they analyze the case of *elastic demand.* Let f_{u_j} be a nondecreasing and concave real function with $f_{u_j}(0) > 1$ for all $u_j \in V$. Then

$$W_F^{\text{bin. el.}}(Y_r|X_p) := \sum_{u_j \in \{u_i \in V | D(u_i, Y_r) < D(u_i, X_p)\}} \frac{\pi(u_j)}{f_{u_j}(D(u_j, Y_r))} \tag{2.8}$$

defines the follower's market share. The authors show that an optimal solution need not exist, so that they seek ϵ-optimal solutions, i.e. solutions that guarantee an objective function value at most ϵ units away from a known upper bound. Furthermore, they show that an ϵ-optimal solution with all of the follower's facilities located at a vertex or sufficiently close to an isodistant point of the network always exists.

Well known results due to Hakimi (1986, 1990) are related to alternative modifications of the basic $(r|X_p)$-medianoid problem. The author presents a *partially binary choice rule with inelastic demand*, where f_{u_j} is a nondecreasing and concave real function with $f_{u_j}(0) > 0$ for all $u_j \in V$, i.e.

$$W_F^{\text{part. bin. inel.}}(Y_r|X_p) := \sum_{u_j \in V} \frac{\pi(u_j) f_{u_j}(D(u_j, X_p))}{f_{u_j}(D(u_j, X_p)) + f_{u_j}(D(u_j, Y_r))}, \tag{2.9}$$

a *partially binary choice rule with elastic demand*, where f_{u_j} is a nondecreasing and concave real function with $f_{u_j}(0) > 1$ for all $u_j \in V$, i.e.

$$W_F^{\text{part. bin. el.}}(Y_r|X_p) := \sum_{u_j \in V} \frac{\pi(u_j)(f_{u_j}(D(u_j, X_p)) - 1)}{\left[f_{u_j}(D(u_j, X_p)) f_{u_j}(D(u_j, Y_r)) - 1\right]}, \tag{2.10}$$

a *proportional choice rule with inelastic demand*, where f_{u_j} is a nondecreasing and concave real function with $f_{u_j}(0) > 0$ for all $u_j \in V$, i.e.

$$W_F^{\text{prop. inel.}}(Y_r|X_p) := \sum_{u_j \in V} \sum_{x_i \in Y_r} \frac{\frac{\pi(u_j)}{f_{u_j}(d(u_j, x_i))}}{\sum\limits_{x_k \in Y_r \cup X_p} \frac{1}{f_{u_j}(d(u_j, x_k))}}, \tag{2.11}$$

and, finally, a *proportional choice rule with elastic demand*, where f_{u_j} is a nondecreasing linear function with $f_{u_j}(0) > 1$ for all $u_j \in V$, i.e.

$$W_F^{\text{prop. el.}}(Y_r|X_p) := \sum_{u_j \in V} \sum_{x_i \in Y_r} \frac{\frac{\pi(u_j)}{f_{u_j}(d(u_j, x_i)) - 1}}{1 + \sum\limits_{x_k \in Y_p \cup X_p} \frac{1}{f_{u_j}(d(u_j, x_k)) - 1}}. \tag{2.12}$$

Hakimi (1986, 1990) proves the existence of an optimal solution Y_r^* with $Y_r^* \subset V$ to the $(r|X_p)$-medianoid problems in all of these cases (cf. also Bauer et al., 1993, for the proportional inelastic case). Note that in the latter two cases Hakimi assumes $p + r \leq n$. For the proportional elastic case, Suárez-Vega et al. (2004a) show that Hakimi's result remains true for f_{u_j} being nondecreasing concave real functions with $f_{u_j}(0) > 1$ for all $u_j \in V$. Moreover, they present computational results on the performance of some heuristic methods that are applied to problem instances of the discretized versions of the basic $(r|X_p)$-medianoid problem as well as its extensions related to formulas (2.8)–(2.12).

Dasci et al. (2002) generalize the basic $(r|X_p)$-medianoid problem by incorporating edge demand as described in Chapter 1.3. Ties are broken equally. Since the basic problem with vertex demand only is NP-hard (see Table 2.3), the generalized version with additional edge demand is obviously NP-hard too. The authors show that the problem is still NP-hard if vertex demands are dropped by reducing the Dominating Set problem (cf. Garey and Johnson, 1979) to the resulting $(r|X_p)$-medianoid problem with edge demand only. Furthermore, they show that a simple augmentation of the discretization result of Megiddo et al. (1983) can be applied to the $(r|X_p)$-medianoid problem with edge (or vertex and edge) demands. This augmentation needs to take into account that the problem might not have an optimal solution, so that they seek ϵ-optimal solutions. One needs to additionally consider the set of all γ-points, i.e. points located γ units away from a vertex, where γ has to be sufficiently small. The authors then derive an algorithm of time complexity $\mathcal{O}(nm^2)$ to solve the $(1|X_p)$-medianoid problem with vertex and edge demand.

Another contribution dealing with a $(1|X_p)$-medianoid problem with edge demand (cf. Chapter 1.3) is Okunuki and Okabe (2002). The authors assume that N is planar and that $\mathcal{O}(p) < \mathcal{O}(n) = \mathcal{O}(m)$. Furthermore, consumers with deterministic, inelastic demand probabilistically choose among facilities. The probabilistic choice rule is based on the (network) *Huff-model.* According to this model, the probability of a customer visiting a facility is proportional to the attractiveness (for example the size) of the facility and inversely proportional to some non decreasing, positive function g_{dist} of the distance to it.[2] Define a_i to be a measure of the attractiveness of facility $i = 1, ..., p+1$ and let $\lambda > 0$. The proposed model of consumer choice then defines the probability for a customer located at point x to visit facility $i \in 1, ..., p+1$ located at x_i as follows:

2 For details on the Huff-model and related customer choice models, see the references in Okunuki and Okabe (2002) or the reviews on consumers in location models by Drezner and Eiselt (2002) and spatial interaction modeling by Roy and Thill (2004).

$$pr_{x,x_i} := \frac{a_i e^{-\lambda d(x,x_i)}}{\sum_{j=1}^{p+1} a_j e^{-\lambda d(x,x_j)}}. \tag{2.13}$$

The expected value of edge customers visiting facility i located at x_i is

$$E(x_i) := \sum_{[u,v]\in E} \int_{x=0}^{\lambda(uv)} pr_{x,x_i}\delta(uv)dx. \tag{2.14}$$

Okunuki and Okabe (2002) then derive an algorithm of time complexity $\mathcal{O}(n^2 \log n)$ to find approximately optimal solutions to the corresponding $(1|X_p)$-medianoid problem. The algorithm is based on transformations of the original network and numerical methods.

The Huff-model (with a concave function g_{dist}) is also applied to a $(1|X_p)$-medianoid problem with vertex demand only by Peeters and Plastria (1998), who point to an anomaly of this proportional choice rule: Customers will patronize very distant facilities even when there exist more attractive facilities that are much closer. The authors therefore propose the usage of a modified Huff-model, the *Pareto-Huff-model*. In this model, the customers located at vertex $u_j \in V$ will only visit a subset of facilities, denoted by $Par_{u_j}(X_p \cup \{x_{p+1}\})$, that are Pareto optimal with respect to the attractiveness and the distance:

$$f \in Par_{u_j}(X_p \cup \{x_{p+1}\})$$
$$\Leftrightarrow \forall g \in X_p \cup \{x_{p+1}\} \backslash \{f\} \begin{cases} \textit{either } a_g < a_f, \\ \textit{or } d(u_j, g) > d(u_j, f), \\ \textit{or } a_g = a_f \\ \textit{and } d(u_j, g) = d(u_j, f). \end{cases} \tag{2.15}$$

Demand is assumed to be deterministic and inelastic. While there always exists an optimal solution at a vertex for the regular Huff-model, the same does not hold for the Pareto-Huff model. Peeters and Plastria (1998) derive a discretization result which states that there is always an optimal solution at, or an ϵ-optimal solution close to, a vertex or an isodistant point of the network for any $\epsilon > 0$. The authors show that there exist at most $2m\hat{n}p$ isodistant points in a network, where $\hat{n}$ is the number of vertices $v \in V$ with $\pi(v) \neq 0$, and propose an $\mathcal{O}(\hat{n}n \log n + pm\hat{n}^2)$ enumeration algorithm to determine all (ϵ-) optimal solutions of the problem. Pelegrín et al. (2010) extend these results by proving that the discretization result holds for the case of multiple facilities, i.e. $r \geq 1$, as well.

Suárez-Vega et al. (2007b) present another model that incorporates attractiveness levels of facilities in a network with vertex demand only. Let $f_{u_j} : \mathbb{R}_0^+ \to \mathbb{R}$ be an increasing and concave function for all $u_j \in V$ and $A_p = (a_1, ..., a_p)$ and $A_r = (a_{p+1}, ..., a_{p+r})$ be the attractiveness levels of the facilities. Furthermore, associate an *attraction threshold* τ_{u_j} with every vertex $u_j \in V$, such that the users at u_j patronize facility i at point $x_i \in N$ with attractiveness level a_i only if $a_i / f_{u_j}(d(u_j, x_i)) \geq \tau_{u_j}$. We denote a subset $S_Z(u_j)$ of a set Z of facility locations that capture demand at node $u_j \in V$ by $S_Z(u_j) := \{x_k \in Z | \frac{a_k}{f_{u_j}(d(u_j,x_k))} \geq \tau_{u_j}\}$. Then

$$W_F^{\text{attr.}}(Y_r, A_r | X_p, A_p) := \sum_{u_j \in \{u_i \in V | S_{Y_r}(u_i) \neq \emptyset\}} \pi(u_j) \frac{\sum_{x_i \in S_{Y_r}(u_j)} \frac{a_i}{f_{u_j}(d(u_j,x_i))}}{\sum_{x_i \in S_{Y_r \cup X_p}(u_j)} \frac{a_i}{f_{u_j}(d(u_j,x_i))}} \tag{2.16}$$

defines the follower's market share. Given an attractiveness level a_i and a vertex $u_j \in V$, a point $x \in N$ is said to be a (u_j, τ_{u_j}, a_i)*-threshold point*, if the condition $f_{u_j}(d(u_j, x)) = a_i / \tau_{u_j}$ holds. Each element of the set

$$T(a_i) = \cup_{u_j \in V} \{x \in N | x \textit{ is a } (u_j, \tau_{u_j}, a_i)\textit{-threshold point}\} \tag{2.17}$$

is called a threshold point with respect to the attractiveness level a_i. Suárez-Vega et al. (2007b) prove that, given X_p, A_p and A_r, there exists an optimal solution Y_r^* to the $(r|X_p)$-medianoid problem under consideration, where every facility $x_i^* \in Y_r^*$ with attractiveness level a_i is a vertex or an element of $T(a_i)$.

While so far the models have considered location decisions based on given attractiveness levels, Suárez-Vega et al. (2004b) define both, locations and attractiveness levels of the follower, as decision variables (cf. also Santos-Peñate et al., 2007). Thus, they define what they call a $(r|X_p, A_p)$*-medianoid problem*: given the attractiveness levels A_p and locations X_p of the leader, the follower rationally chooses attractiveness levels A_r and locations Y_r. Obviously, one needs to include a cost of attractiveness in this case. The authors do so by including fixed cost of attractiveness $F(a_j)$, where F is a positive, continuous and non-decreasing function of a non-negative real variable, and maximizing profit, i.e. market share minus the sum of attractiveness cost for all facilities. They analyze inelastic vertex demand under binary, partially binary and proportional choice rules, i.e. extensions of the basic $(r|X_p)$-medianoid problem and the problems related to (2.9) and (2.11). Suárez-Vega et al. (2004b) provide discretization results for the partially binary and proportional choice rules, while in the case of binary customer choice a discretization can only be applied if attractiveness levels are given. Furthermore, they present some heuristic algorithms and computational results for the resulting discrete problems. Suárez-Vega et al. (2007a) extend these results by providing a discretization result for the $(r|X_p, A_p)$-medianoid problem related to (2.12) (proportional, elastic customer choice), where f_{u_j} are nondecreasing and concave real functions for all $u_j \in V$.

García Pérez et al. (2000) (cf. also García Pérez and Pelegrín Pelegrín, 1997) examine a $(1|X_1)$-medianoid problem on a tree network with vertex customers only. The follower's potential locations are restricted to a given path within the tree network. Demand is assumed to be inelastic and customers select the closest facility (binary choice). The entrant is assumed to have two main objectives, the first is to maximize market share, the second is to minimize the maximum travel time of the customers that are being served by the entrant. This *bicriteria optimization model* might have applications in situations where the time to attend a demand is important, as, for example, in messenger delivery services or fast food services. The authors assume travel time to be proportional to distance and show that the problem on a tree network can be reduced to the same type of problem on a chain network.

They define a location to be efficient, if there exists no other location with equally large or larger market share and equally large or smaller maximum travel time. However, since the infimum value of the maximum travel time might not be reached for some of the provably finite number of values of possible market shares of the follower, García Pérez et al. (2000) consider ϵ-efficient locations, that are defined in line with ϵ-optimality, as well. Finally, the authors derive an algorithm to determine the entire set of efficient and ϵ-efficient locations.

García et al. (2011) incorporate prices into a $(r|X_p)$-medianoid problem on a network.[3] As, in most markets, the choice of location is less flexible than the choice of prices, locational models usually assume that simultaneous *price competition* occurs after the location decisions have been made. Thus, García et al. (2011) allow the players to change prices in a simultaneous pricing game (Betrand game) after the follower's entry. The follower anticipates the price competition in this separate stage when making the location decisions. The players are assumed to be profit maximizers and employ spatial price discrimination. Customers are located at the vertices of the network and patronize the facility that offers the lowest price (binary choice). The price sensitive demand at a vertex $u_j \in V$ of the network is described by a continuous and strictly decreasing function $q_{u_j}(p)$ of the price $p \in [0, p_{u_j}^{max}]$, where $p_{u_j}^{max}$ is the reservation price of the users at u_j. Note that spatial price discrimination is applied to $(r|X_p)$-medianoid problems with constant vertex demand by Pelegrín et al. (2006) (for $r = 1$) and Pelegrín et al. (2010) as well. García et al. (2011) assume that the players are able to price below $p_{u_j}^{max}$ for each $u_j \in V$. Furthermore, they assume that the marginal delivered cost (the sum of marginal cost of production and transportation) at each $u_j \in V$ is independent of the amounts sold and that the players will not price below their marginal delivered cost. Therefore, a separate Betrand game takes place for every vertex of the network for any fixed set of locations. García et al. (2011) show that each player will monopolize the set of vertices

3 Note that García et al. (2011) assume that multiple players are initially located in the network. These players may be interpreted as a single opponent from the point of view of the entrant. Thus, we may interpret the problem setting as a $(r|X_p)$-medianoid problem.

where his marginal delivered cost is lower than the minimum marginal delivered cost of the other players. The players then either set the optimal monopoly price or the price equal to the minimum marginal delivered cost of the competitors in each of their monopolized vertices and a price equal to their minimum marginal delivered cost in the other vertices. Thus, the location price problem is reduced to a location problem if all players are assumed to set these equilibrium prices. The authors then additionally assume the marginal production cost to be a positive concave function on an edge of the network and the marginal transportation cost to be a positive, concave and increasing function with respect to the distance of a facility to a vertex of the network. They prove that, under these assumptions, there exists a set of vertices which is an optimal solution to the $(r|X_p)$-medianoid problem.

We close this section by extending Table 2.3 in listing complexity results that have been explicitly mentioned in the surveyed papers in Table 2.4.

Tab. 2.4: Some complexity results on selected $(r|X_p)$-medianoid problems

Problem characteristics	NP-hard	Polynomial
(vertex and) edge demand binary choice (closest facility) deterministic, inelastic demand	$(r\|X_1)$ on general networks (Dasci et al., 2002)	$(1\|X_p)$: $\mathcal{O}(nm^2)$ ϵ-opt. solution on general networks (Dasci et al., 2002)
edge demand prop. choice (Huff model) deterministic, inelastic demand		$(1\|X_p)$: $\mathcal{O}(n^2 \log n)$ approx. opt. solution on general networks (Okunuki and Okabe, 2002)
vertex demand Pareto-Huff model deterministic, inelastic demand		$(1\|X_p)$: $\mathcal{O}(\hat{n}n \log n + pm\hat{n}^2)$ ϵ-opt. solution on general networks (Peeters and Plastria, 1998)

2.4 (r|p)-Centroid Problems

Hakimi (1983) was the first to publish complexity results on the *basic $(r|p)$-centroid problem*, which we define in line with the basic $(r|p)$-medianoid problem, i.e. we consider vertex customers who patronize the closest facility to accommodate their deterministic, inelastic demand. He proved the basic $(1|p)$-centroid problem on a general network to be NP-hard by reduction of the Vertex Cover problem (cf. Garey and Johnson, 1979). Another well known result is due to Hansen and Labbé (1988) who provided an algorithm of polynomial time complexity to determine the set of all optimal solutions to the basic $(1|1)$-centroid problem on general networks (cf. already Section 2.2).

These complexity results have recently been extended by Spoerhase and Wirth (2009c). The authors provide a NP-hardness proof for the basic $(r|p)$-centroid problem on chain networks that uses a reduction of the Partition problem (cf. Garey and Johnson, 1979). Furthermore, they investigate the basic $(1|p)$-centroid problem. They show that Hakimi's result on its complexity on general networks remains true on networks of bounded pathwidth (pw-bounded networks), while a discretization result can be utilized to design an exact algorithm of polynomial time complexity on tree networks. Note that the authors assume $\pi : V \rightarrow \mathbb{Q}_0^+$, $\lambda : E \rightarrow \mathbb{Q}^+$ and $\lambda([u,v]) = d(u,x) + d(x,v)$ for any point x on an edge $[u,v] \in E$. The complexity results are summarized in Table 2.5.

Tab. 2.5: Some complexity results on a basic $(r|p)$-centroid problem

Problem characteristics	NP-hard	Polynomial
customers located at vertices, binary choice (closest facility), deterministic demand, inelastic demand	$(1\|p)$ on general networks (Hakimi, 1983) $(1\|p)$ on pw-bounded networks (Spoerhase and Wirth, 2009c) $(r\|p)$ on chain networks (Spoerhase and Wirth, 2009c)	$(1\|1)$: $\mathcal{O}(n^4m^2\log(mn)\log(\pi(N)))$ on general networks (Hansen and Labbé, 1988) $(1\|p)$: $\mathcal{O}(n^2(\log n)^2\log\pi(N)\log\hat{D})$ on tree networks (Spoerhase and Wirth, 2009c)

Based on Table 2.5 and Section 2.3, one may conclude that centroid problems are substantially harder than medianoid problems (cf. also Section 2.6). This is supported by Hakimi (1983), who defines an *approximate solution* to a $(r|p)$-centroid problem as a solution with an objective function value that is within a fixed positive (integer) factor α of being optimal, where α is not a function of the size of the network. He proves that even the problem of finding an approximate solution to the basic $(1|p)$-centroid problem is NP-hard. Spoerhase and Wirth (2008) extend this result by showing that there exists a fully polynomial time approximation scheme for the basic $(r|p)$-centroid problem on chain networks with $\pi : V \to \mathbb{N}$ and $\lambda : E \to \mathbb{N}^+$.

Motivated by the fact that the basic $(r|p)$-centroid problem is polynomially time solvable on tree networks for $r = p = 1$, research in this field has focused mainly on incorporating additional features into this special case. Shiode and Drezner (2003), for example, extend the basic $(1|1)$-centroid problem on a tree network by assuming that the vertex weights at the follower's time of entry used by the leader in his decision are *stochastic*; that is, they are random variables according to mutually independent normal distributions $\pi(u_j) \sim N(\mu_j, \sigma_j^2)$ with mean μ_j and variance σ_j^2 for all $u_j \in V$. The follower then locates his facility knowing the location of the leader and the vertex demands. Note that the authors do not permit co-location, i.e. locating at the same point of the network, of the two players. The leader wishes to minimize the maximum total demand taken away by the follower with a satisfying probability β with $0.5 < \beta < 1$. It is easy to see that there always exists an optimal location of the follower at a vertex of the tree network. Shiode and Drezner (2003) additionally prove that there always exists an optimal solution to this stochastic $(1|1)$-centroid problem at a vertex of the tree network and present a simple efficient solution procedure based on an enumeration of the vertices and bisection.

An alternative extension to the basic $(1|1)$-centroid problem on tree networks is presented by García Pérez and Pelegrín Pelegrín (2003). The authors incorporate parametric prices, i.e. they assume that the unit mill price of the product is fixed for each player. Transportation cost is assumed to be a linear function of distance. The demand at each node is served by the lower cost facility (parametric price + transportation cost). Moreover, the authors assume that each player gets a positive profit once located at an optimal location, so that competition is possible. Hereafter, they present two $\mathcal{O}(n^3 \log n)$ algorithms to generate the set X_1^* of all optimal leader locations under the assumption that the parametric prices are different. Note that the algorithm presented by Spoerhase and Wirth (2009b) (cf. Section 2.2) may be applied to compute the set X_1^* in $\mathcal{O}(n \log n)$ time. García Pérez and Pelegrín Pelegrín (2003) determine this set for equal parametric prices as well. Furthermore, they prove that there always exists an optimal location for the follower at a vertex of the tree network when prices are different, while this does not have to be the case for the leader. When prices are equal, there always exists an optimal location for the leader at a vertex, while this might not be true for the follower.

García Pérez and Pelegrín Pelegrín (1997) analyze the basic (1|1)-centroid problem on a tree network for different objective functions. They, first, consider the maximization of market share and, second, the minimization of the maximum travel time of the customers. They prove that, if we restrict the players' locations to a path within the tree network, the game can be reduced to a corresponding game on a chain network and that the optimal locations of both, the leader and the follower, can be determined analytically.

While García Pérez and Pelegrín Pelegrín (1997) study the variation of objective functions that affect both players, Berman and Gavious (2007) analyze a two player leader follower game on a network with differing objectives of the players. The leader, a state, locates p facilities that provide support in case of a terrorist attack. Locating an emergency facility incurs a fixed cost $C > 0$. Moreover, the state decides on investing resources c_{prev} on prevention, such that the state strategy is a vector $s = (x_1, ..., x_p, c_{prev})$. The follower, a terrorist, attacks one of the cities, modeled as vertices of the network. For the case of more than one leader facility the authors analyze discrete versions of the problem, while results for a continuous version are presented for the case of one facility ((1|1)-centroid problem). The vertex weights of the graph are interpreted as the expected damage in case of a terrorist attack on that city. Edge weights represent the delay of shipment of resources from the facility. The terrorist may use a mixed strategy $t = (p_1, ..., p_n)$ where p_i is the probability that the terrorist will attack vertex $u_i \in V$. It is assumed that $\sum_{i=1}^{n} p_i = 1$ and $p_i \geq 0$ for all i. The probability for an attack to succeed is modeled by a continuous decreasing convex function $P(c_{prev})$ with $P(0) > 0$ and $\lim_{c_{prev} \to \infty} P(c_{prev}) = 0$. The disutility in case of a successful terrorist attack on city u_i is

$$f_i := (\alpha d(x,i) + \eta_i)\pi(u_i), \tag{2.18}$$

where x is the closest facility to the city and $\alpha > 0$ represents the cost of delaying one unit of resource due to a unit of distance. Furthermore, $\eta_i := \eta + \alpha d(i)$ with $\eta > 0$ and $d(i)$ representing the delay at the city itself. Thus, the expected utility of the leader is described by

$$U_S(t,s) := -P(c_{prev}) \sum_{i=1}^{n} p_i f_i - pC - c_{prev}. \tag{2.19}$$

The terrorist's utility is defined to be

$$U_T(t,s) := \sum_{i=1}^{n} p_i(\gamma d(x,i) + \delta_i) w_i \tag{2.20}$$

with $\delta_i := \delta + \alpha d(i)$ and $\delta > 0$, $\gamma > 0$. Berman and Gavious (2007) show that an optimal solution to the $(1|1)$-centroid problem is included in a finite set of points, called local centers, that can be determined analytically. Moreover, they show that an optimal solution can be determined efficiently.

2.5 Multiple Players, Endogenous Location and Some Selected $\mathbb{R}^1$ Models

So far, we have been able to interpret most of the games using Hakimi's framework of $(r|X_p)$-medianoid and $(r|p)$-centroid problems with problem specific assumptions and extensions. This is no longer reasonable if we consider more complex games with, for example, more than two players or endogenized market entry or location order. These kinds of assumptions, as suggested by Eiselt and Laporte (1996) for future research, have mainly been applied to linear markets. Table 2.6 lists some selected recent contributions.

Granot et al. (2010) allow an endogenously determined, potentially indefinite number of players to locate one facility on a line and, in an extension to this model, a network that may contain loops. Players locate in the location space, at fixed cost, according to an exogenously given order as long as it is profitable for them to do so. The demand of the uniformly spread customers is assumed to be elastic and modeled by a demand intensity function, denoted by $d(x)$, which is continuous and decreasing in the distance x to the closest facility (binary choice). Thus, in the linear model (and analogously in the network model) the revenue of a player, located at x_0, from the costumers in a subinterval $(x_0, x_0 + r)$, provided he is the closest player to these customers, is

$$D(r) := \int_0^r d(x)dx. \tag{2.21}$$

Prices are fixed to 1. Making some assumptions with regard to the demand intensity function and not permitting co-location, the authors show in their main result for the network model that, in equilibrium, players will first locate at the vertices of the network according to some easily computable order, and subsequently, they will locate on the edges. This result can be generalized to the case of edge specific demand intensity functions. Moreover, Granot et al. (2010) prove that the optimal number of facilities to be located by a monopolist is strictly smaller than the number of players who will be located in equilibrium of the competitive model if edge lengths are not too small.

Other features that have mainly been considered in sequential competitive location models on a line include asymmetric information (concerning, for example, customer preferences or the cost structure of the opponent), asymmetric (direction dependent) transportation cost, special settings in the pricing game (concerning the timing of events) and players with different objectives (global or local players, public or private firms, etc.). We list some related publications in Table 2.7.[4] The reader may find interesting ideas to be adapted in general network models in future.

4 Note that some of the authors analyze extensions to their models while we refer to their main models in the Table 2.7.

Tab. 2.6: Selected models on $\mathbb{R}^1$ with multiple players or endogenous location order

		Players		Pricing and cost			Customers		
Authors	Space	Number	Loc. order	Pricing	Setup/prod. cost	Transp. cost	Distribution	Demand	Choice
Aiura (2010)	$[-1,1]$	3	exogenous	none	none	linear	uniform on unknown interval of length 1, multi period model with updated beliefs	inelastic	binary
Götz (2005)	$[0,1]$	endogenous	exogenous	mill price set simult.	fixed setup cost	quadratic	uniform with variable density	inelastic	binary
Harter (1997)	$[0,2]$	n (exogenous)	exogenous	mill price set simult.	none	quadratic	uniform on unknown interval of length 1	inelastic	binary
Heywood and Ye (2009a,b)	$[0,1]$	n (exogenous), different "types"	exogenous	spatial price discrimination	none	linear	uniform	inelastic	binary
Lambertini (1997)	$\mathbb{R}^1$	2	endogenous (simult. entry possible)	mill price, timing endogenous	none	quadratic	uniform on interval of length 1	inelastic	binary
Loertscher and Muehlheusser (2008)	two $[0,1]$ markets	3, different "types"	exogenous	none	none	linear	uniform with market specific density	inelastic	binary
Meza and Tombak (2009)	$[0,l]$, $l \in \mathbb{R}^+$	2	endogenous (simult. entry possible)	mill price set simult.	marginal prod. cost asymmetries	quadratic	uniform	inelastic	binary
Nilssen (1997)	$[0,1]$	3	exogenous	none	none	linear, direction dependent	uniform	inelastic	binary
Rothschild et al. (2007)	$[0,1]$	3, strategic contracts, merging	exogenous	spatial price discrimination	none	linear	uniform	inelastic	binary
Yates (1997)	$[0,1]$	n (exogenous)	exogenous	none	none	linear, direction dependent	uniform	elastic	binary

Refer to the caption underneath Table 2.7 for an illustration of the cell entries.

Tab. 2.7: Other selected $\mathbb{R}^1$ models (continues on next pages)

		Players		Pricing and cost			Customers		
Authors	Space	Number	Loc. order	Pricing	Setup/prod. cost	Transp. cost	Distribution	Demand	Choice
Beladi et al. (2010)	$[0,1]$	2, vertically related	exogenous	spatial price discrimination	none	linear	uniform	inelastic or elastic (multiple products)	binary
Bonein and Turolla (2009)	$\mathbb{R}^1$	2	exogenous	mill price set simult.	none	quadratic	uniform on interval of length 1, asymmetric information	inelastic	binary
Boyer et al. (2003a) (cf. also Boyer et al., 1995)	$[0,l]$, $l \in \mathbb{R}^+$	2	exogenous	mill price set simult.	marginal. prod. cost asymmetries, asymmetric information	quadratic	uniform	inelastic	binary
Boyer et al. (2003b) (cf. also Boyer et al., 1995)	$[0,l]$, $l \in \mathbb{R}^+$	2	exogenous	mill price set simult. or spatial price discrimination	marginal. prod. cost asymmetries, asymmetric information, fixed setup cost	linear, quadratic	density fctn.	inelastic	binary
Dasci and Laporte (2005, 2007)	$[0,1]$	2, endogenous number of facilities	exogenous	none	fixed setup cost	linear	density fctn.	inelastic	binary
Fleckinger and Lafay (2010)	$[0,1]$	2	exogenous	mill price set at the same time as loc. decision	none	general fctn.	density fctn.	inelastic	binary
Lai (2001)	$[0,1]$	2	exogenous	none	none	linear, direction dependent	uniform	elastic	binary
Lambertini (2002)	$[0,1]$	2, uncertainty over follower's time of entry	exogenous	mill price set simult.	none	quadratic	uniform	inelastic	binary

Tab. 2.7: (continued)

		Players		Pricing and cost			Customers		
Authors	Space	Number	Loc. order	Pricing	Setup/prod. cost	Transp. cost	Distribution	Demand	Choice
Matsumura and Matsushima (2003)	$[0, 1]$	2, different "types"	exogenous	mill price set simult., price regulation	none	quadratic	uniform	inelastic	binary
Matsumura and Matsushima (2010)	$[0, 1]$	2	exogenous	mill price set simult.	marginal prod. cost can be changed by investment in a pregame stage	quadratic	uniform	inelastic	binary
Nilssen and Sørgard (2002)	$[0, 1]$	2, different "types"	exogenous	none	none	linear, direction dependent	uniform	elastic	binary
Peng and Tabuchi (2007a,b)	$[0, 1]$	2, endogenous number of facilities, endogenous number of product varieties	exogenous	parametric price	fixed cost per variety	quadratic	uniform	elastic	binary
Rhee (2006)	$[0, 1]$	2	exogenous	mill price set simult.	none	quadratic	uniform, customers differ along unobservable characteristics	inelastic	binary
Tabuchi and Thisse (1995)	$\mathbb{R}^1$	2	exogenous	mill price set simult.	none	quadratic	uniform or triangular on $[0, 1]$	inelastic	binary
Tyagi (1999)	$\mathbb{R}^1$	2	exogenous	mill price, endogenous order	none	quadratic	uniform on bounded interval	inelastic	binary
Tyagi (2000)	$\mathbb{R}^1$	2	exogenous	mill price set simult.	marginal prod. cost asymmetries	quadratic	uniform on $[-0.5, 0.5]$	inelastic	binary

Tab. 2.7: (continued)

		Players		Pricing and cost			Customers		
Authors	Space	Number	Loc. order	Pricing	Setup/prod. cost	Transp. cost	Distribution	Demand	Choice
Zhou and Vertinsky (2001)	$[0,1]$	2	exogenous	mill price set simult., monopoly price before follower's entry, interest rate	fixed setup cost (may be different for players)	quadratic	uniform, time dependent density	inelastic	binary

Columns:

space	specification of location space
number	information on number of players; endogenously or exogenously determined; "types", e.g. public or private, domestic or foreign firms; contracts among firms; vertically related industries; product varieties (refer to the papers for details)
loc. order	location order endogenously or exogenously determined
pricing	type of pricing subgame
setup/prod. cost	information on setup or production cost
transp. cost	information on transportation cost; linear or quadratic function of distance; dependent on or independent of direction of travel
distribution	type of user distribution
demand	elastic or inelastic demand
choice	choice rule

2.6 On the Influence of Specific Modeling Assumptions

As pointed out by Eiselt and Laporte (1996), sequential competitive location problems are very sensitive with respect to the underlying modeling assumptions, especially in terms of complexity, the existence and nature of solutions and discretization results. D'Aspremont et al. (1979), for example, show that there is no Nash equilibrium in pure strategies in Hotelling's (first location then price) game with linear transportation cost.[5] However, by assuming the transportation cost to increase quadratically with the distance, the authors are able to derive a proof for the existence of a Nash equilibrium in which the competitors locate at the opposite ends of the line segment. Other examples are related to voting theory: In Section 2.1 we have illustrated that the existence of Condorcet points (and related concepts) often depends on the structure of the underlying network and the distribution of the users. This led to the incorporation of minimax objectives (Section 2.2) or a threshold value α. Sections 2.3–2.5, however, have mainly been describing the modeling assumptions themselves in detail. It is the aim of this section to explore the above-mentioned sensitivity of these models. In doing so, we assume the reader to be familiar with the effects summarized in Eiselt and Laporte (1996). Moreover, we note that the papers that have been subject of Sections 2.3 and 2.4 are mainly concerned with existence, complexity and discretization issues, while research on $\mathbb{R}^1$ models (Section 2.5) concentrates on effects like agglomeration, maximal differentiation, leader and follower advantages and the existence of equilibria. Therefore, this section is divided into two parts, highlighting the implications of general network models and models on a line separately.

Consider an arbitrary $(r|p)$-centroid problem on a network. Due to the sequential nature of the game, we are (in mathematical programming terms) faced with a bilevel (or hierarchical) optimization problem (see Bard, 1998; Dempe, 2002, for an introduction to bilevel programming). The fact that

5 Note that Hotelling (1929) claimed that, in equilibrium, the two competitors will agglomerate in the center of the line segment.

even the linear bilevel programming problem in continuous variables is NP-hard in the strong sense (Bard, 1998, chap. 5) suggests, that we are unlikely to be confronted with polynomially time solvable problems. Indeed, as Table 2.5 demonstrates, the basic $(r|p)$-centroid problem is NP-hard even if we consider relatively simple network structures as, for example, chain networks. By fixing $r = 1$, one can derive an efficient algorithm on tree networks; A feature which is lost when considering a slightly more complex network class: pathwidth bounded networks. A polynomial time algorithm for the basic $(r|p)$-centroid problem on general networks is only known for the case $r = p = 1$. Thus, two of the major factors influencing the complexity of centroid problems are the network structure and the number of facilities to be located by the players.[6] Tables 2.3 and 2.4 show that this result carries over to the basic $(r|X_p)$-medianoid problem if we treat r as an arbitrary input parameter. On tree networks, we are able to efficiently determine an optimal solution for arbitrary values of r, while this is not the case on general networks. Furthermore, the influence of the number of facilities on the complexity persists when considering additional features as, for example, edge demand (Dasci et al., 2002) or other types of choice rules (Okunuki and Okabe, 2002).

It is well known that modeling assumptions influence the existence and cardinality of finite dominating sets of competitive location problems. This, in turn, may effect the applicability of enumerative algorithms or the neighborhood definitions of potential heuristic algorithms. In the case of $(r|X_p)$-medianoid problems (see Section 2.3), such factors include:

- Binary choice: While there exists a subset of the vertex set (frequently referred to as the "Hakimi property" or "vertex optimality") that corresponds to an optimal solution in the case of choice rules (2.9)–(2.12) (partially binary and proportional choice with inelastic and elastic demand, respectively), finite dominating sets are usually substantially larger in the case of binary choice rules. This is true for both, inelas-

6 The dependency of complexity on network structure is well known in the (competitive and non-competitive) location literature. Kincaid (2011) gives an overview of classical contributions in this field.

tic and elastic demand (basic $(r|X_p)$-medianoid problem and choice rule (2.8)). Furthermore, it remains true when considering $(r|X_p, A_p)$-medianoid problems under different choice rules (Suárez-Vega et al., 2007a).

- Pareto optimality: The same effect arises when considering (ϵ-optimal solutions of) the Pareto-Huff-model instead of the Huff-model (Peeters and Plastria, 1998).
- Edge demand: When (additionally) incorporating edge demand, one may need to seek ϵ-optimal solutions (Dasci et al., 2002). The resulting effect on the cardinality of finite dominating sets is rather small.

Discretization results concerning $(r|p)$-centroid problems (Section 2.4) are rather limited. A basic $(r|1)$-centroid of a general network is always a vertex for $r \geq 2$, while this is not the case for $r = 1$. On tree networks, however, there always exists a vertex that is a $(1|1)$-centroid (see Hakimi, 1990, and the references therein). The latter statement remains true in the case of stochastic demand as presented by Shiode and Drezner (2003), but it may be wrong when parametric prices are included (García Pérez and Pelegrín Pelegrín, 2003). An additional discretization result for the case $r = p = 1$ is presented by Berman and Gavious (2007). Furthermore, Spoerhase and Wirth (2009c) derive a (non polynomial) discretization result for the basic $(1|p)$-centroid problem on a tree network.

As mentioned above, modeling assumptions do not only influence the complexity and discretization results of the underlying problem, but may also effect the existence and nature of its solutions. In Section 2.3, for example, due to the potential non-existence of optimal solutions, ϵ-optimal solutions had to be considered in the case of choice rule (2.8), when considering edge demand or when applying the Pareto-Huff model. Results concerning $\mathbb{R}^1$ models are listed in Table 2.8. Those models are typically designed with regard to analytic traceability or analyzed by simulation. Complexity issues are usually not considered in detail.

Tab. 2.8: $\mathbb{R}^1$ *- Main features and results (continues on next pages)*

Authors	Main features	Results
	Demand uncertainty	
Aiura (2010)	Demand uncertainty in a multi period model: Customers distributed on an unknown interval of length 1. Later entrants may adjust predictions on the distribution by observing demand signals arising after preceding firms enter the market.	Agglomeration of the three firms occurs (only) if firms can observe signals on the customer distribution. In this case they dislike revealing the signals and therefore chose the same location as the preceding firm.
Bonein and Turolla (2009)	Customers distributed on an interval of length 1. One player is perfectly informed while the other faces demand uncertainty.	Demand uncertainty leads to differentiation when faced by the leader while it is an agglomeration force when faced by the follower. Welfare losses are higher in the latter case.
Harter (1997)	Customers distributed on an unknown interval of length 1.	Except from the duopoly case, firms locate almost evenly throughout the location space.
	Customer attributes	
Götz (2005)	Uniform customer distribution with variable density (market size).	Locational patterns are in general asymmetric in the duopoly case. Profits are non-monotonic in market size. Equilibrium profits of all firms may be larger in situations in which more firms are active.
Rhee (2006)	Consumers differ both in their locations and in their tastes along unobservable characteristics (random component in customer's utility functions).	The more the customer choices depend on the unobservable characteristics, the closer to the center of the market the follower locates. Anticipating this behavior, the leader locates farther away from the center. Not only the locations, but also the degree of first mover advantage (or disadvantage) is determined by the level of dependence of the customer behavior on unobservable characteristics.

Tab. 2.8: (continued)

Authors	Main features	Results
Zhou and Vertinsky (2001)	Time dependent customer density. Inclusion of interest rate. Leader monopoly prior to follower's entry.	The follower always maximally differentiates. The leader chooses the center of the market if transportation cost and market growth rate are sufficiently low and interest rate and fixed cost are sufficiently large. Otherwise the leader will locate as far away from the center as possible.
	Different "types" of players	
Beladi et al. (2010)	Vertically related industry with one upstream and two downstream firms who move sequentially. The latter produce two out of three differentiated goods each.	A merged firm (one of the downstream firms and the upstream firm) will locate farther away from the social optimum than the firm outside the merger.
Heywood and Ye (2009a,b)	Sequential games with a domestic public firm, domestic private firms and foreign firms. Domestic welfare is defined to be consumer surplus plus domestic profit and equals global welfare minus the profit of the foreign firm.	Without considering the entry of a foreign firm, the presence of a public firm generally increases welfare. The entry of a foreign firm often lowers domestic welfare. Privatization of the public firm may lower domestic welfare but can increase global welfare.
Loertscher and Muehlheusser (2008)	A global player competes with local players on two (heterogeneous) linear markets with the same product.	Agglomeration occurs across markets.
Matsumura and Matsushima (2003)	Sequential games with a public and a private firm. The effect of price regulation is considered.	The desirable role (leader or follower) of the public firm depends on the presence of price regulation. Neither price regulation nor privatization of the public firm improves welfare.
Rothschild et al. (2007)	Possibility of strategic contracts or merging between two out of three firms that enter sequentially.	The outcomes in terms of welfare consequences and location choice are substantially different according to different modes of entry.

Tab. 2.8: (continued)

Authors	Main features	Results
	Endogenous timing	
Lambertini (1997)	The timing of entry and pricing is endogenously determined. Players may locate outside the consumer interval.	The (subgame perfect) equilibrium in pure strategies is reached by playing sequentially in the price stage after having set locations simultaneously.
Meza and Tombak (2009)	Asymmetric firms with different marginal production cost. The timing of entry is endogenously determined.	The existence of an equilibrium in pure strategies, the timing of events and the locations themselves depend on the marginal cost differences. Usually, sequential market entry naturally arises, the low cost firm takes the role of the leader and the follower maximally differentiates.
	Pregame investments	
Matsumura and Matsushima (2010)	There is an ex ante marginal production cost difference (efficient leader, inefficient follower). Marginal production cost can be changed by investment in a pregame stage.	An ex ante inefficient follower has a stronger incentive to invest, given that the ex ante cost difference is sufficiently large. A decrease in cost of this follower pushes the facilities apart, resulting in increasing prices and market share of the follower. Decreasing cost of the ex ante efficient leader lead to decreasing differentiation and decreasing prices.
	Asymmetric transportation cost	
Lai (2001)	Transportation cost is finite only in one direction of travel.	No (subgame perfect) equilibrium in pure strategies.
Nilssen (1997)	Asymmetric transportation cost.	Locational patterns (as well as a potential first mover advantage) depend on the explicit values of the transportation costs. There exists no indifference among late movers that want to locate between early movers.

Tab. 2.8: (continued)

Authors	Main features	Results
Nilssen and Sørgard (2002)	Asymmetric transportation cost. Public leader, private follower.	Both, triggered differentiation and triggered agglomeration, may be welfare maximizing (depending on the asymmetry of transportation costs).
Yates (1997)	Transportation cost is finite only in one direction of travel.	Equilibria are symmetric in the sense that all players obtain the same market share. Note that, in contrast to Lai (2001), Yates (1997) considers ϵ-optimal locations.
	Information asymmetry	
Boyer et al. (2003a,b)	Information asymmetry concerning marginal production cost.	Information asymmetry may not cause any distortion in the location game if a fixed cost of entry is not present. But, in general, information plays a crucial role in location strategies preventing a follower's entry.
Lambertini (2002)	An infinite time horizon is included. The follower's time of entry is uncertain. As long as the follower has not entered, the leader acts as a monopolist.	The later the follower is expected to enter, the closer to the center of the market the leader locates.
Tyagi (2000)	Marginal production cost asymmetries. Players may locate outside the consumer interval.	Unless the leader is certain that the follower will not have a superior cost structure, it may be better to locate away from the market center.
	Special pricing game	
Fleckinger and Lafay (2010)	The players (simultaneously or sequentially) choose price and location at the same time. A general transportation cost function is applied.	There is no equilibrium in the simultaneous game while there exist equilibria in the sequential case. The follower charges a higher price and always earns more than the leader. Differentiation is never minimal (agglomeration) nor maximal.

Tab. 2.8: (continued)

Authors	Main features	Results
Tyagi (1999)	The order of setting prices after the location stage is endogenized. Players may locate outside the consumer interval.	The unique equilibrium outcome involves players choosing sequential pricing over simultaneous pricing. The leader in location acts as the leader in prices.
	Location outside the consumer interval	
Tabuchi and Thisse (1995)	Triangular consumer density. Players may locate outside the consumer interval.	For a symmetric triangular density, only asymmetric equilibria (in pure strategies) exist (even for simultaneous choice of locations). A first mover advantage is induced by allowing location outside the consumer interval.
	Location densities	
Dasci and Laporte (2005, 2007)	The players open an endogenously determined (fixed cost) number of facilities. Strategies are defined in terms of their location densities (no precise representation of the locations).	The model is solvable analytically. If the follower enters the market, he opens at least as many facilities as the leader. The leader can effectively deter entry (even in the case of a cost disadvantage). However, this may not always be rational. Entry and entry deterrence decisions are quite sensitive to consumer densities and fixed cost. This is not the case for the location strategies, once both players are present.
	Product varieties	
Peng and Tabuchi (2007a,b)	Competition in location and product varieties. Endogenization of the number of facilities and varieties of the players by including a fixed cost per variety.	If each player is restricted to locating a single facility, neither agglomeration nor maximum differentiation occur. Allowing multiple stores yields a rich set of equilibrium outcomes.

2.7 Conclusion

In this chapter we have presented a review of recent developments in the field of sequential competitive location problems with a focus on problems defined on networks. We have included the class of voting location problems. In the latter context, we have given an overview of different kinds of point sets in networks, as, for example, generalizations of the well known Condorcet and Simpson sets, and their interrelationships. We have outlined the features and main results of recently published $(r|X_p)$-medianoid and $(r|p)$-centroid problems. Moreover, we have presented a tabular overview of more complex games that have mainly been defined on linear markets. These models feature, for example, an endogenized market entry or location order, multiple players, asymmetric information, asymmetric transportation cost or different player objectives. Finally, we have outlined the effects of some modeling assumptions on the complexity, the existence and nature of solutions and discretization results of the problems under consideration. We refer to Younies and Eiselt (2011) for ideas on future research directions.

Chapter 3

Competitive or Voting Location with Proportional Choice

This chapter is concerned with a competitive or voting location problem on networks under a proportional choice rule that has previously been introduced by Bauer et al. (1993) (cf. already Sections 2.1 and 2.2). Its contribution is threefold:

- First, as in Kress and Pesch (2012b), we refine a discretization result of Bauer et al. (1993) by proving convexity and concavity properties of related expected payoff functions and we answer the long time open question whether 1-suboptimal points are always vertices by providing a counterexample on a tree network. These results are presented in Section 3.2.1 and are complemented by examples and counterexamples in Section 3.2.2.
- Second, in Section 3.2.3, we are concerned with implementational issues of an algorithm for determining 1-suboptimal points of a network.
- Third, we will show that the choice rule of Bauer et al. (1993) can be extended to include edge demand in Section 3.3.

In order to be able to define the proportional choice rule and its properties in detail, however, we will first have to introduce the concept of bottleneck points in Section 3.1 (cf. also Hakimi, 1964; Hooker et al., 1991). The chapter closes with a conclusion and remarks on future research in Section 3.4.

3.1 Bottleneck Points

Let $N = (V, E, \lambda)$ be a finite network. Furthermore, let $[u, v] \in E$ be any edge and $i \in N$, $i \notin (u, v)$ be a point of the network (see Figure 3.1a). Define

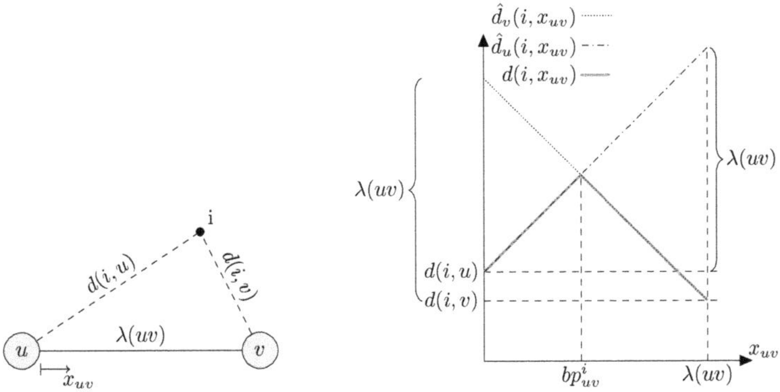

(a) Edge and point *(b) Graphical interpretation*

Fig. 3.1: Bottleneck points

$\hat{d}_u(i, x_{uv})$ (and $\hat{d}_v(i, x_{uv})$) to be the length of the – not necessarily shortest – path from a point $x_{uv} \in [0, \lambda(uv)]$ to i via vertex u (or v) using the shortest path from u (or v) to i, that is $\hat{d}_u(i, x_{uv}) := d(i, u) + x_{uv}$ (and $\hat{d}_v(i, x_{uv}) := d(i, v) + \lambda(uv) - x_{uv}$). If there exists a point $x_{uv} = bp^i_{uv} \in (0, \lambda(uv))$ such that $\hat{d}_u(i, bp^i_{uv}) = \hat{d}_v(i, bp^i_{uv})$, then $bp^i_{uv} = 0.5(d(i, v) + \lambda(uv) - d(i, u))$ (see Figure 3.1b). Such a point is generally referred to as a *bottleneck point* of the edge $[u, v]$ with respect to point i and defines the most remote point from i on edge $[u, v]$. The bottleneck point is defined to be one of the vertices u or v in the following cases:

$$bp^i_{uv} := \begin{cases} 0 & \text{if } d(i,u) = d(i,v) + \lambda(uv), \\ \lambda(uv) & \text{if } d(i,v) = d(i,u) + \lambda(uv). \end{cases} \tag{3.1}$$

We get

$$d(i, x_{uv}) = \begin{cases} d(i,u) + x_{uv} & \text{if } 0 \le x_{uv} \le bp^i_{uv}, \\ d(i,v) + \lambda(uv) - x_{uv} & \text{if } bp^i_{uv} < x_{uv} \le \lambda(uv). \end{cases} \tag{3.2}$$

Now assume $i \in (u,v)$, i.e. i is a point on edge $[u,v]$, $i \neq u$, $i \neq v$. We define two bottleneck points

$$bp_{uv}^{i,left} := \begin{cases} \frac{1}{2}(2i - d(u,v) - \lambda(uv)) & \text{if } i \geq \frac{1}{2}(\lambda(uv) + d(u,v)), \\ 0 & \text{else,} \end{cases} \tag{3.3}$$

and

$$bp_{uv}^{i,right} := \begin{cases} \frac{1}{2}(2i + d(u,v) + \lambda(uv)) & \text{if } i \leq \frac{1}{2}(\lambda(uv) - d(u,v)), \\ \lambda(uv) & \text{else.} \end{cases} \tag{3.4}$$

Note that at least one of the bottleneck points $bp_{uv}^{i,left}$ and $bp_{uv}^{i,right}$ coincides with a vertex of the corresponding edge. Figure 3.2 illustrates this fact by providing three examples with $d(u,v) < \lambda(uv)$ and $i < 0.5(\lambda(uv) - d(u,v))$ (Figure 3.2a), $0.5(\lambda(uv) - d(u,v)) < i < 0.5(\lambda(uv) + d(u,v))$ (Figure 3.2b) and $i > 0.5(\lambda(uv)+d(u,v))$ (Figure 3.2c). The bottleneck points are depicted as gray dots.

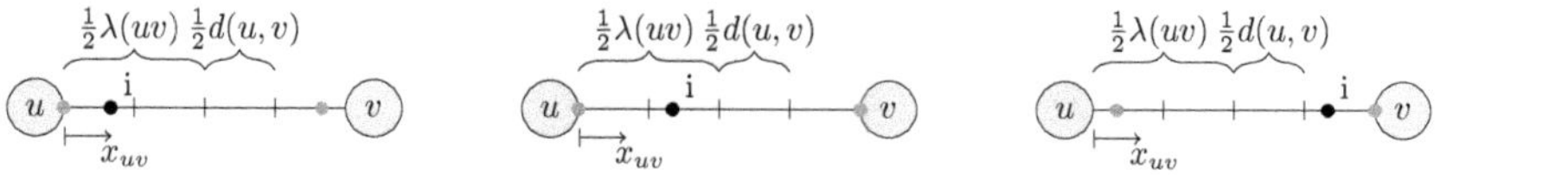

(a) $bp_{uv}^{i,right} < \lambda(uv)$ *(b)* $bp_{uv}^{i,left} = 0$, $bp_{uv}^{i,right} = \lambda(uv)$ *(c)* $bp_{uv}^{i,left} > 0$

Fig. 3.2: Bottleneck points $bp_{uv}^{i,left}$ *and* $bp_{uv}^{i,right}$

It follows that

$$d(i, x_{uv}) = \begin{cases} x_{uv} + d(u,v) + \lambda(uv) - i & \text{if } 0 \leq x_{uv} < bp_{uv}^{i,left}, \\ \lambda(uv) - x_{uv} + d(u,v) + i & \text{if } bp_{uv}^{i,right} < x_{uv} \leq \lambda(u,v), \\ |x_{uv} - i| & \text{else.} \end{cases} \tag{3.5}$$

Now observe that, given an arbitrary edge $[u,v] \in E$ and any point i of the network, the function $d(i, x_{uv})$ is continuous in the variable x_{uv} on the interval $[0, \lambda(uv)]$. This is easily seen from (3.2) and (3.5), since an increase of a given x_{uv} by an arbitrarily small $\epsilon > 0$ (with $x_{uv}+\epsilon \leq \lambda(uv)$) can change the distance $d(i, x_{uv})$ by at most ϵ.

3.2 A Proportional Choice Rule

Bauer et al. (1993) consider proportional choice in the context of a single facility voting location problem with vertex customers (vertex voters) only. The facility to be located in this problem corresponds to a leader's facility in the corresponding (two player) sequential competitive location problem. In order to make the distinction of leader and follower more intuitive, we will refine the notation and definitions of Bauer et al. (1993). This induces the need to revisit some of their propositions and theorems.

Let $N = (V, E, \lambda)$, $V = \{u_1, ..., u_n\}$, be a (finite) network and let $x \in N$ be the leader's location and $y \in N$ the follower's location, respectively.[1] We assume that the customers as well as the commodity offered by the players are homogenous. Then we define

$$E_L^k(x, y) := \begin{cases} \sum_{i=1}^{n} \pi(u_i) \cdot \frac{d(u_i,y)^k}{d(u_i,y)^k + d(u_i,x)^k} & \text{if } x \neq y, \\ \frac{1}{2}(\pi(V) + \pi(x)) & \text{if } x = y, \end{cases} \tag{3.6}$$

to be the expected number of customers visiting the leader's (index L) facility for any given $k \in \mathbb{N} \cup \{\infty\}$.[2] k may be interpreted as a parameter for customer loyalty, since a very large k models the deterministic (binary) behavior of the customers in the Condorcet model (cf. Bauer et al., 1993). Similarly, the expected number of customers accommodating their demand at the follower's (index F) facility is defined to be

1 For the sake of notational convenience we will occasionally denote the vertices by natural numbers, i.e. i instead of u_i.

2 Note that $E_q^k(x, y)$, $q \in \{L, F\}$, does not refer to the edge set E of the underlying network and that k is an index with respect to $E_q^k(x, y)$, while it is an exponent of the distances. Furthermore, note that we will usually drop the index k when $k = 1$ in the remainder of the chapter.

$$E_F^k(x,y) := \begin{cases} \sum_{i=1}^{n} \pi(u_i) \cdot \frac{d(u_i,x)^k}{d(u_i,x)^k + d(u_i,y)^k} & \text{if } x \neq y, \\ \frac{1}{2}\left(\pi(V) - \pi(x)\right) & \text{if } x = y. \end{cases} \tag{3.7}$$

It is easy to see that we have $E_L^k(x,y) + E_F^k(x,y) = \pi(N)$. Note that, if the follower's facility y is established at the same site as the leader's facility x, ties are broken such that all customers $v \in V \setminus \{x\}$ (are expected to) accommodate half of their demand at each of the facilities, while the customers located at x (are expected to) visit the leader's facility only. This assumption is designed to avoid trivial solutions (similar to Hakimi, 1990) and to guarantee continuity of the expected value functions (Proposition 3.2.1, cf. Bauer et al., 1993). However, it can also be motivated from a practical point of view by interpreting it as enforcing a kind of local first mover advantage. That is, the leader may have superior information on the local market that can, for instance, be utilized for effective sales promotion.

Proposition 3.2.1. *Let N be a (finite) network. The expected value functions $E_L^k(x,y)$ and $E_F^k(x,y)$ are continuous in y on N. Furthermore, assuming $y \notin V$, they are continuous in x on N.*

Proof. The proof is an immediate consequence of Propositions 1 and 2 of Bauer et al. (1993). An alternative proof may make use of the continuity observation at the end of Section 3.1. □

Observe that, given some $y \in V$ with $\pi(y) > 0$, $E_L^k(x,y)$ and $E_F^k(x,y)$ have exactly one discontinuity at $x = y$. It is easy to see that this is a result of the tie breaking rule given above.

In analogy to the Condorcet and Simpson concepts we define:

Definition 3.2.1 (Bauer et al. (1993)). Let N be a (finite) network. Given two points $x, y \in N$, the value $E_F^k(x,y)/\pi(N)$ is the *relative k-rejection* of point x at point y. The *maximal relative k-rejection* of a point $x \in N$ is the value

$$\rho^k(x) := \max_{y \in N} \frac{E_F^k(x,y)}{\pi(N)}.$$

Definition 3.2.2 (Bauer et al. (1993)). Let N be a (finite) network. A point $x \in N$ is called *k-optimal* if

$$E_L^k(x,y) \geq \frac{1}{2}\pi(N)$$

for all $y \in N$. A point $x \in N$ is called *k-suboptimal* if

$$\rho^k(x) = \min_{z \in N} \rho^k(z).$$

A main result of Bauer et al. (1993) states that k-optimal points are always vertices:

Theorem 3.2.1 (Bauer et al. (1993)). *Let N be a network with vertex weight function $\pi : V \to \mathbb{R}_0^+$ and $\pi(N) > 0$. Let $I(u,v)$ be a ported interval of N where $\pi(I(u,v)) = \pi(\{u,v\})$ and let x be any point of this interval where $u \neq x \neq v$. Then, x is not a k-optimal point for each $k \in \mathbb{N}$.*

Based on this theorem and Proposition 3.2.4 (see below) the authors derive an algorithm of time complexity $\mathcal{O}(n^3)$ to determine all 1-optimal points of a network.

3.2.1 1-Suboptimal Points

The following convexity property will prove useful in the sequel.

Proposition 3.2.2. *Let N be a (finite) network and $\hat{x} \in V$ be a distinct leader's location. The expected values $E_F(\hat{x}, x_{uv})$ are convex functions in the variable x_{uv} on $I = [0, \lambda(uv)]$ for all $[u,v] \in E$.*

Proof. Given $\pi(u_i) \geq 0$ and Proposition 3.2.1, it is sufficient to show that the functions

$$e_{u_i,F}(\hat{x}, x_{uv}) := \frac{d(u_i,\hat{x})}{d(u_i,\hat{x}) + d(u_i, x_{uv})}$$

are convex in x_{uv} on the open interval $\bar{I} = (0, \lambda(uv))$ for all $\hat{x}, u_i \in V$ and $[u,v] \in E$. It is easy to see that the functions $e_{u_i,F}(\hat{x}, x_{uv})$ are convex (and non-increasing) in the (non-negative) distance $d(u_i, x_{uv})$ (a formal proof can be found in Hakimi, 1986). Now, given the previously derived continuity observation and the fact that $d(u_i, x_{uv})$ is the minimum of two linear functions in x_{uv} (see Section 3.1), the assertion follows immediately. □

Similarly, we find:

Proposition 3.2.3. *Let N be a (finite) network and* $\hat{y} \in V$ *be a distinct follower's location. The expected values* $E_F(x_{uv}, \hat{y})$ *are concave functions in the variable* x_{uv} *on* $I = [0, \lambda(uv)]$ *for all* $[u, v] \in E$.

Proof. The proof is in analogy to the proof of Proposition 3.2.2. □

As a consequence of Proposition 3.2.2 and Theorem 5 of Bauer et al. (1993) we may conclude:[3]

Proposition 3.2.4. *Let N be a network with vertex weight function* $\pi : V \to \mathbb{R}_0^+$ *and at least two positive vertex weights. Let* $I(u, v)$ *be a ported interval of N where* $\pi(I(u, v)) = \pi(\{u, v\})$ *and let y be any point of this interval where* $u \neq y \neq v$. *Then* $E_F(x, u) > E_F(x, y)$ *or* $E_F(x, v) > E_F(x, y)$ *for all* $x \in N$.

Thus, the maximal relative 1-rejection of any point of a network is always reached at a vertex.

Definition 3.2.3. Let N be a (finite) network and $k = 1$. We say R is the *upper envelope of rejection*, if $R(x_{uv}) = \max_{w \in V} E_F(x_{uv}, w)$ for any $x_{uv} \in [0, \lambda(uv)]$ and for all $[u, v] \in E$.[4]

Given Propositions 3.2.3 and 3.2.4, it is easy to see that any 1-suboptimal point is at a vertex of the network N or at a point $x \in N$ where "rejection functions" that define the upper envelope of rejection intersect, i.e. where $E_F(x, w_1) = E_F(x, w_2) = R(x)$, $w_1, w_2 \in V$, $w_1 \neq w_2$. Making use of

3 Note that, in their proof of Theorem 5 under the assumptions stated in Proposition 3.2.4, Bauer et al. (1993) essentially only show that $E_F(x, u) > E_F(x, y)$ or $E_F(x, v) > E_F(x, y)$ for all $x \in N \setminus \{u, v\}$ (not for all $x \in N$). Proposition 3.2.2 fixes this inaccuracy.

4 We will occasionally use the term "upper envelope of relative rejection" which refers to the upper envelope of rejection that has been scaled by dividing by $\pi(V)$.

this discretization property, Bauer et al. (1993) design an algorithm of time complexity $\mathcal{O}(n^5)$ to determine all 1-suboptimal points of a given network (cf. Section 3.2.3 for details). It remained open if 1-suboptimal points are even always vertices. An example answers this long time open question for the case of tree networks. Consider the network of Figure 3.3 with all edge lengths equal to one and vertex weights $\pi(v) \in \{0,1\}$ for all $v \in V$, the latter being indicated by numbers next to the vertices.

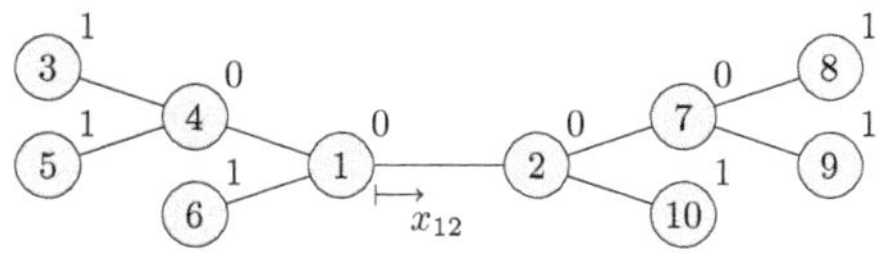

Fig. 3.3: Tree network

Table 3.1 lists the relative 1-rejection of vertex $i \in V$ (leader's location) at vertex j (follower's location) in cell (i,j), when considering the example network. Thus, when restricting the set of potential locations to the vertex set, the maximal relative 1-rejection is minimal at vertices 1 and 2. We denote this level of rejection by ρ_0.

Tab. 3.1: Relative 1-rejection of vertices

	1	2	3	4	5	6	7	8	9	10
1	0.5	0.5	0.472	0.487	0.472	**0.5095**	0.486	0.462	0.462	0.486
2	0.5	0.5	0.462	0.487	0.462	0.486	0.487	0.472	0.472	**0.5095**
3	0.528	0.538	0.417	0.492	0.5	0.514	0.528	0.5	0.5	0.514
4	0.513	0.514	0.508	0.5	0.508	0.5	0.5	0.472	0.472	0.49
5	0.528	0.538	0.5	0.492	0.417	0.514	0.528	0.5	0.5	0.514
6	0.49	0.514	0.486	0.5	0.486	0.417	0.5095	0.486	0.486	0.5
7	0.514	0.513	0.472	0.5	0.472	0.49	0.5	0.508	0.508	0.5
8	0.538	0.528	0.5	0.528	0.5	0.514	0.492	0.417	0.5	0.514
9	0.538	0.528	0.5	0.528	0.5	0.514	0.492	0.5	0.417	0.514
10	0.514	0.49	0.486	0.5095	0.486	0.5	0.5	0.486	0.486	0.417

Figure 3.4 takes a closer look at edge [1, 2] and corresponding rejection functions $E_F(x_{12}, w)$, $w \in V$. Note that, for the sake of clarity, only the subset of functions that define the upper envelope of rejection is depicted. We find that every point on edge [1, 2] has a maximal relative 1-rejection level smaller than or equal to ρ_0. As a direct consequence we may conclude that

there exists no 1-suboptimal point at a vertex of the network. Furthermore, when analyzing the remaining edges of the network, we find two points on edge $[1, 2]$, so_1 and so_2, to be 1-suboptimal.

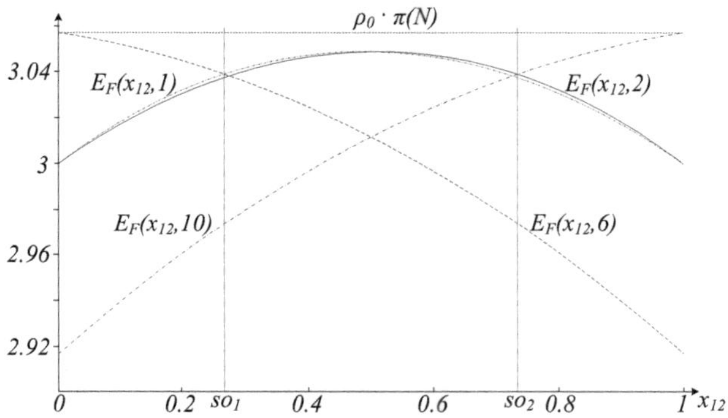

Fig. 3.4: Rejection functions and 1-suboptimal points

We close this section by providing a vertex optimality property for the family of connected networks with two vertices only.

Proposition 3.2.5. *Let* $N = (V, E, \lambda)$, $V = \{u, v\}$, $E = \{[u, v]\}$ *be a network with vertex weight function* π *and at least one positive vertex weight. Then* $\rho^k(u) < \rho^k(x_{uv})$ *or* $\rho^k(v) < \rho^k(x_{uv})$ *for all* $x_{uv} \neq 0$ *and* $x_{uv} \neq \lambda(uv)$.

Proof. It is easy to see that $\rho^k(u) = \frac{\pi(v)}{\pi(N)}$ and $\rho^k(v) = \frac{\pi(u)}{\pi(N)}$. Therefore,

- if $\pi(u) = \pi(v)$, we have $\rho^k(u) = \rho^k(v) = 0.5$,
- if $\pi(u) > \pi(v)$, we have $\rho^k(u) < 0.5$,
- if $\pi(u) < \pi(v)$, we have $\rho^k(v) < 0.5$.

Thus, there always exists a k-optimal point, which, in combination with Theorem 3.2.1, i.e. the fact that k-optimal points are always vertices, proves the assertion. □

3.2.2 Examples and Counterexamples

When analyzing the example of Figure 3.3 in Section 3.2.1, the reader may have had several ideas on varying the network to achieve the vertex optimality property or simplifying the network without enforcing vertex optimality. This section aims at providing counterexamples on some of these potential ideas that may also raise questions for future research.

First, we show that the vertex optimality property may not hold even when $\pi(v) > 0$ for all $v \in V$. To this end, we augment the example network by introducing a sufficiently small constant $\epsilon > 0$ and adding this constant to all zero vertex weights (see Figure 3.5, where $\epsilon_1 = \epsilon_2 = \epsilon$). The relevant rejection functions for $\epsilon = 0.02$ are depicted in Figure 3.6a (all elements are in analogy to Figure 3.4). It is easy to see that the vertex optimality property does not hold. Future research may analyze how small the differences in vertex weights in general networks can get without enforcing vertex optimality.

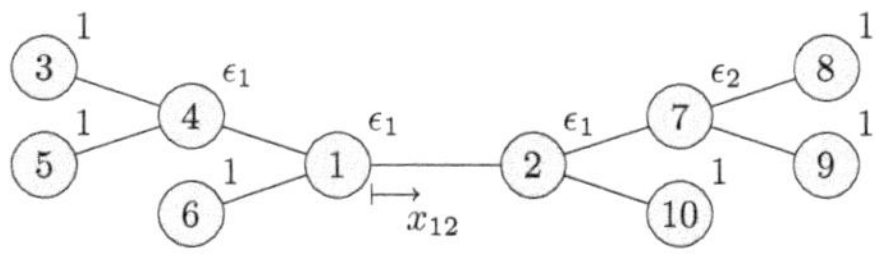

Fig. 3.5: No vertex weight equals zero

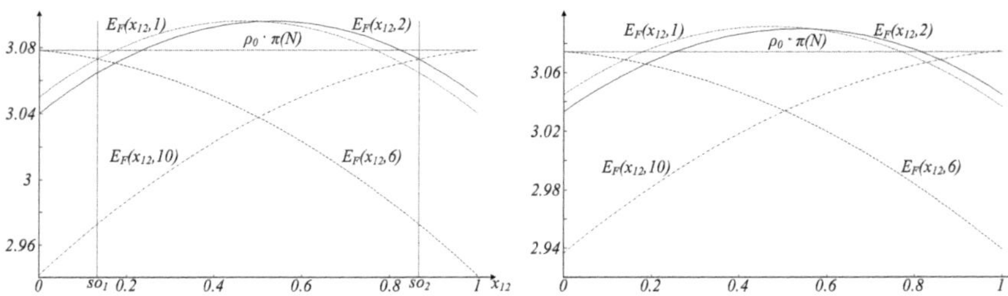

(a) $\epsilon_1 = \epsilon_2 = \epsilon = 0.02$ (b) $\epsilon_1 = 0.02$, $\epsilon_2 = 0.01$

Fig. 3.6: Rejection functions and 1-suboptimal points

We may conclude from the former example that there exist networks $N = (V, E, \lambda)$ that do not possess the vertex optimality property, even if $\pi : V \to \mathbb{N}^+$, i.e. $\pi(v) > 0$ and integer for all $v \in V$. To see this, suppose that all edge lengths and vertex weights of a network $N = (V, E, \lambda)$ are rational numbers. Represent those numbers as fractions of two integers and define c to be the least common multiple of their denominators. Then we can transform N into a network $N' = (V, E, \lambda')$ with $\lambda'(uv) = c\lambda(uv)$ for all $[u, v] \in E$ and $\pi'(u) = c\pi(u)$ for all $u \in V$ without changing the ratio of the (expected) market shares of leader and follower for any feasible location setting.

Next, one may wonder if the symmetry of the example network plays a crucial role. Thus, we break the symmetry by setting $\epsilon_1 = 0.02 \neq \epsilon_2 = 0.01$ in the network of Figure 3.5. This results in the rejection functions shown in Figure 3.6b. Again, the vertex optimality property does not hold.

We say that two neighbored vertices $v_1, v_2 \in V$ of a network $N = (V, E, \lambda)$ are being merged, if the corresponding edge length $\lambda(v_1, v_2)$ is reduced to zero, which is equivalent to introducing a new vertex v_{v_1,v_2} with $\pi(v_{v_1,v_2}) = \pi(v_1) + \pi(v_2)$ in the place of vertices v_1 and v_2 and edge $[v_1, v_2]$. Table 3.2 lists merging operations that the reader may have thought of in order to simplify the example network of Figure 3.3 without inducing vertex optimality. Observe that a 1-optimal point exists in all of the resulting networks, and thus any 1-suboptimal point is a vertex (cf. Theorem 3.2.1).

Tab. 3.2: Merging vertices

merge vertices	3,5	3,5 and 8,9	3,4	3,4 and 7,9
1-optimal points	$v_{3,5}$	$v_{3,5}$ and $v_{8,9}$	$v_{3,4}$	$v_{3,4}$ and $v_{7,9}$
merge vertices	6,1	6,1 and 2,10	3,5 and 7,9	3,5 and 2,10
1-optimal points	$v_{6,1}$	$v_{6,1}$ and $v_{2,10}$	$v_{3,5}$ and $v_{7,9}$	$v_{3,5}$
merge vertices	3,4 and 2,10	3,5 and 6,1	3,5; 6,1; 2,10 and 8,9	
1-optimal points	$v_{2,10}$	$v_{3,5}$	$v_{6,1}$ and $v_{2,10}$	

In analogy to Table 3.2, Table 3.3 refers to deleting vertices and their incident edges in the example network. Again, a 1-optimal point exists in all of the resulting networks.

Tab. 3.3: Deleting vertices

delete vertices	3	6	3,5	3,6	3,8	3,10	6,10	3,5,8
1-optimal points	10	10	8,9	8,9	6,10	8,9	3,5,8,9	10
delete vertices	3,5,10	3,6,8	3,6,10	3,5,8,9	3,5,8,10	3,6,8,10		
1-optimal points	8,9	10	8,9	6,10	6,9	5,9		

3.2.3 Determining 1-Suboptimal Points: Implementational Issues

Given an integer parameter p and a sufficiently small $\epsilon \in \mathbb{R}^+$, Bauer et al. (1993) present the following method for determining all 1-suboptimal points of a network N with vertex demand only and at least two positive vertex weights.

Algorithm 3.2.1 Determine all 1-suboptimal points of a network N

$P := \emptyset \quad \forall e \in E,\ E_{cand} := E$
$\rho_0 := \min_{u \in V} \rho(u)$ (Bauer et al. (1993) provide an $\mathcal{O}(n^3)$ algorithm)
while $E_{cand} \neq \emptyset$ **do**
  repeat
    choose any subedge (of an edge $[u, v]$) $x_1 y_1 \in E_{cand}$[5]
    $E_{cand} := E_{cand} \setminus \{x_1 y_1\}$
  until $\rho(x_1) \leq \rho_0$ **or** $\rho(y_1) \leq \rho_0$
  $k := 1$
  while $|d(u, x_k) - d(u, y_k)| > \epsilon$ **and** $|P| < n + 2$ **do**
    let $w_k \in V$ be a vertex such that $E_F(x_k, w_k) = \rho(x_k)$
    define the points λ_k und μ_k by
      $d(u, \lambda_k) := d(u, x_k) + (F_{p-k-1}/F_{p-k+1}) d(x_k, y_k)$
      $d(u, \mu_k) := d(u, x_k) + (F_{p-k}/F_{p-k+1}) d(x_k, y_k)$
    $\rho_k := \min\{\rho(x_k), \rho_{k-1}\}$
    if $\rho(x_k) < \rho_{k-1}$ **then**
      $P := \{x_k\}$
    else if $\rho(x_k) = \rho_{k-1}$ **then**
      $P := P \cup \{x_k\}$
    end if
    if $E_F(\lambda_k, w_k) < \rho_k > E_F(\mu_k, w_k)$ **then**
      $E_{cand} := E_{cand} \cup \{\lambda_k y_k\}$
      $x_{k+1} := x_k,\ y_{k+1} := \lambda_k$
    else if $E_F(\lambda_k, w_k) \geq \rho_k > E_F(\mu_k, w_k)$ **then**

---

5 At this point of the algorithm, Bauer et al. (1993) assume that $x_{uv} = 0$ at vertex u and that $x_1 \leq y_1$.

$x_{k+1} := \lambda_k,\ y_{k+1} := y_k$
else if $E_F(\mu_k, w_k) \geq \rho_k > E_F(y_k, w_k)$ **then**
$x_{k+1} := \mu_k,\ y_{k+1} := y_k$
end if
if $|d(u, x_{k+1}) - d(u, y_{k+1})| > \epsilon$ **then**
exchange x_{k+1} and y_{k+1}
$k := k + 1$
else
$\rho_0 := \rho(\alpha_k)$ where $d(u, \alpha_k) := \frac{1}{2}(d(u, x_k) + d(u, y_k))$
if $\rho_0 < \rho_k$ **then**
$P := \{\alpha_k\}$
else if $\rho_0 = \rho_k$ **then**
$P := P \cup \{\alpha_k\}$
end if
end if
end while
if $|d(u, x_k) - d(u, y_k)| > \epsilon$ **then**
$E_{cand} := E_{cand} \cup \{x_k y_k\}$
$\rho_0 := \rho_k$
end if
if $|P| \geq n + 2$ **then**
$P := \emptyset$
end if
end while

Algorithm 3.2.1 is based on the Fibonacci search method (F_i, $i = 0, 1, \ldots$, correspond to the Fibonacci numbers) and Propositions 3.2.3 and 3.2.4. Basically, as described in Section 3.2.1, it numerically determines intersection points of "rejection functions" that define the upper envelope of rejection and checks their potential of being 1-suboptimal. We refer to the original publication for references and a detailed explanation. However, there are some implementational issues that are subject of the remainder of this section.

Determining Maximal Rejection Values

Bauer et al. (1993) do not provide a procedure to determine the maximal relative 1-rejection of a given point. Making use of the concept of bottleneck points as defined in Section 3.1, an algorithm of time complexity $\mathcal{O}(n^2)$ can easily be derived.[6]

Algorithm 3.2.2 Determine the maximal relative 1-rejection of a point $x_{uv} \in [0, \lambda(uv)]$

```
ρ_abs := 0, A := ∅
for all i := 1 to n do
  ρ_frac := 0
  if x_uv = 0 and i=u then
    ρ_frac := 0.5(π(N) − π(u))
  else if x_uv = λ(uv) and i=v then
    ρ_frac := 0.5(π(N) − π(v))
  else
    for all j := 1 to n do
      if x_uv ≤ bp^j_uv then
        ρ_frac := ρ_frac + π(j) · (d(j,u)+x_uv) / (d(j,u)+x_uv+d(j,i))
      else
        ρ_frac := ρ_frac + π(j) · (d(j,v)+λ(uv)−x_uv) / (d(j,v)+λ(uv)−x_uv+d(j,i))
      end if
    end for
  end if
  if ρ_frac > ρ_abs then
    ρ_abs := ρ_frac, A := {i}
  else if ρ_frac = ρ_abs then
    A := A ∪ {i}
  end if
end for
```

6 Note that we assume the relevant bottleneck points, i.e. the bottleneck points bp^i_{uv} of a given edge $[u, v]$ with respect to the vertices $i \in V$ of the network, to be known. It is easy to see that an algorithm for determining those bottleneck points is of time complexity $\mathcal{O}(n)$.

23: $\rho_{rel} := \frac{\rho_{abs}}{\pi(N)}$

The algorithm determines the maximal relative 1-rejection of a given point in ρ_{rel} by exploiting Proposition 3.2.4. Additionally, it determines the set A of all vertices, in which this rejection level is reached.

Repeat-Until Block (Lines 4–7)

Consider the fictitious example of Figure 3.7 with $w_1, w_2 \in V$ and the depicted relative rejection functions that are assumed to define the upper envelope of relative rejection over the subedge x_1y_1 of edge $[u, v]$.

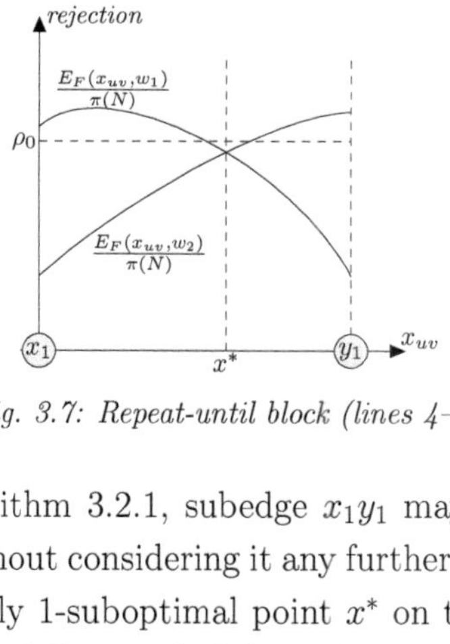

Fig. 3.7: Repeat-until block (lines 4–7)

When calling Algorithm 3.2.1, subedge x_1y_1 may be removed from the candidate set E_{cand} without considering it any further (lines 4–7) even though there exists a potentially 1-suboptimal point x^* on the subedge. Hence, we suggest substituting the following (sub-) procedure instead:

Algorithm 3.2.3 Choose subedge

---

repeat
    $b := true,\ c := false$
    choose any subedge (of an edge $[u, v]$) $x_1y_1 \in E_{cand}$
    $E_{cand} := E_{cand} \setminus \{x_1y_1\}$
    for all w_k with $E_F(x_1, w_k)/\pi(N) = \rho(x_1)$ **do**
        if $E_F(x_1, w_k)/\pi(N) \geq \rho_0$ **and** $E_F(y_1, w_k)/\pi(N) \geq \rho_0$ **then**
            $b := false$
            if $E_{cand} = \emptyset$ **then**

```
            b := true, c := true
          end if
          break (leave for-loop)
        end if
    end for
until b
if c then
    stop algorithm
end if
```

The procedure removes subedges from the candidate set that are guaranteed to not contain any 1-suboptimal point.[7] If, at one point of the modified Algorithm 3.2.1, all remaining subedges are removed ($E_{cand} = \emptyset$), the algorithm terminates.

Exchange x_{k+1} and y_{k+1} (Line 29)

Consider the situation depicted in Figure 3.8 that arises at one point of Algorithm 3.2.1 after having processed line 30, i.e. after having exchanged x_2 and y_2 and having increased k from 1 to 2. We assume $d(u, v) = \lambda(uv)$.

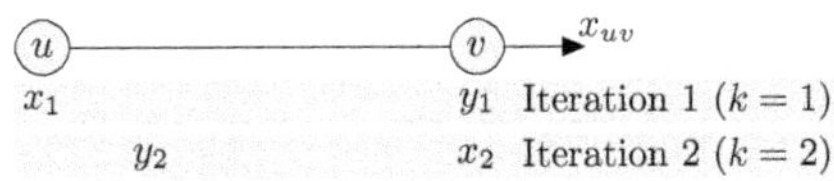

Fig. 3.8: Exchange x_{k+1} and y_{k+1}

When returning to line 12, we have $d(u, x_2) = \lambda(uv)$. Hence, $d(u, \lambda_2) = \lambda(uv) + d(x_2, y_2)(F_{p-2-1} \backslash F_{p-2+1}) > \lambda(uv)$, i.e. λ_2 lies "outside" of the subedge under consideration. To avoid situations like these, we will have to additionally change the direction of counting of the subedge's coordinate (one may interpret this as additionally "exchanging" u and v; see also Figure 3.9) in line 29.

7 A simple modification of Procedure 3.2.3 may additionally take account of potential discontinuities of the rejections functions at vertices.

Case Differentiation (Lines 20–27)

In the case differentiation of lines 20–27, Algorithm 3.2.1 determines subedges that can be dropped from the candidate set E_{cand}. Given Proposition 3.2.3 and the fact that (by definition of the Fibonacci sequence) $\lambda_k < \mu_k$ and (due to line 14) $\rho_k \leq \rho(x_k)$ for all $1 \leq k \leq p-1$, the complete case differentiation is as follows:

1. $E_F(\lambda_k, w_k)/\pi(N) < \rho_k$
   No subedge can be dropped from the candidate set.
2. $E_F(\lambda_k, w_k)/\pi(N) \geq \rho_k$
   (a) $E_F(y_k, w_k)/\pi(N) < \rho_k$
      i. $E_F(\mu_k, w_k)/\pi(N) \geq \rho_k$
         Subedge $[x_k, \mu_k]$ can be dropped from the candidate set.
      ii. $E_F(\mu_k, w_k)/\pi(N) < \rho_k$
         Subedge $[x_k, \lambda_k]$ can be dropped from the candidate set.
   (b) $E_F(y_k, w_k)/\pi(N) \geq \rho_k$
      Subedge $[x_k, y_k]$ can be dropped from the candidate set.

Figure 3.9 illustrates an example for case 2b: First, in iteration k (Figure 3.9a), subedge $[x_k, \lambda_k]$ is dropped from the candidate set, and we set $x_{k+1} = \lambda_k$ and $y_{k+1} = y_k$. Then, we exchange x_{k+1} and y_{k+1} (as well as the corresponding edge's direction of counting) and determine w_{k+1}, λ_{k+1} and μ_{k+1} (Figure 3.9b). Case 2b arises. Note, however, that this situation is not considered in Algorithm 3.2.1 so that the original procedure needs to be modified.

Choosing Parameter p

When implementing Algorithm 3.2.1, one needs (locally or globally) set parameter p to a specific value so that, for every subedge under examination, it is guaranteed that $|d(u, x_k) - d(u, y_k)| \leq \epsilon$ (line 9) at the latest in iteration

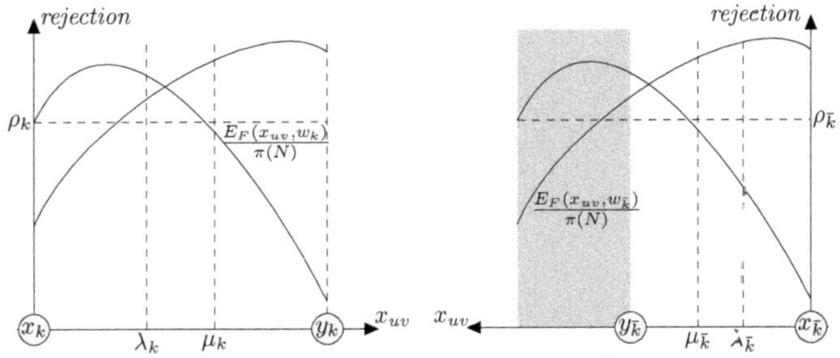

(a) *Iteration k (case 2a.ii)* (b) *Iteration $\bar{k} := k+1$ (case 2b)*

Fig. 3.9: Illustration of the missing case in Algorithm 3.2.1

$k = p - 1$.[8] Consider the case differentiation of the previous subsection (cf. also Bazaraa et al., 2006). We have

$$y_{k+1} - x_{k+1} = \begin{cases} \frac{F_{p-k-1}}{F_{p-k+1}}(y_k - x_k) & \text{for cases 1 and 2a.i,} \\ \frac{F_{p-k}}{F_{p-k+1}}(y_k - x_k) & \text{for case 2a.ii,} \\ 0 & \text{for case 2b.} \end{cases}$$

Hence, after $k = p-1$ iterations and in the worst case, we have $y_p - x_p = (y_1 - x_1)/F_p$. We require $y_p - x_p \leq \epsilon$. Hence, when globally fixing p, we need to choose p so that $\hat{D}/F_p \leq \epsilon$. When choosing p in this manner, one can additionally combine lines 40–46 to a single if statement.

Modified algorithm

We close this section by presenting a modification of Algorithm 3.2.1 based on the implementational issues that have been presented. Note that we explicitly remove all edges $[u, v] \in E$ with $\lambda(uv) > d(u, v)$ from E_{cand} (cf. lines 12 and 13 of the original algorithm). As there are assumed to be at least two positive vertex weights, it is easy to see that such edges can not contain 1-suboptimal points other than in their end points. Observe that we may take account of

8 Alternatively, we may set p to an arbitrary value and leave the (inner) while-loop when $p - k - 1 < 0$.

edges that have been removed from the candidate set when choosing p. We additionally ensure to make use of relative 1-rejection values when necessary and we add an index to P to clarify that 1-suboptimal points can exist on every edge of the network. Let $[x, y]$ by any subedge, then we denote the corresponding edge in E by e_{xy}. Furthermore, before removing subedges from the candidate set as described in Algorithm 3.2.3, we additionally check if their end points are 1-suboptimal.[9]

Algorithm 3.2.4 Determine all 1-suboptimal points of a network N (modified)

$P_e := \emptyset \quad \forall e \in E$, $\rho_0 := \min_{u \in V} \rho(u)$, $E_{cand} := E$
remove all edges $[u, v] \in E$ with $\lambda(uv) > d(u, v)$ from E_{cand}
choose p as described above
while $E_{cand} \neq \emptyset$ **do**
    repeat
        $b := true$, $c := false$
        choose any subedge (of an edge $[u, v]$) $x_1 y_1 \in E_{cand}$
        $E_{cand} := E_{cand} \setminus \{x_1 y_1\}$
        for all w_k with $E_F(x_1, w_k)/\pi(N) = \rho(x_1)$ **do**
            if $E_F(x_1, w_k)/\pi(N) \geq \rho_0$ **and** $E_F(y_1, w_k)/\pi(N) \geq \rho_0$ **then**
                $b := false$
                if $E_{cand} = \emptyset$ **then**
                    $b := true$, $c := true$
                end if
                if $E_F(x_1, w_k)/\pi(N) = \rho_0$ **then**
                    $P_{e_{x_1 y_1}} := P_{e_{x_1 y_1}} \cup \{x_1\}$
                end if
                if $\rho(y_1) = \rho_0$ **then**
                    $P_{e_{x_1 y_1}} := P_{e_{x_1 y_1}} \cup \{y_1\}$
                end if
                break (leave for-loop)

9 Consider, for example, a network with $[a, b], [b, c] \in E$ and $deg(b) = 2$. If both edges, $[a, b]$ and $[b, c]$, were to be removed from E_{cand} in Algorithm 3.2.3, one may not detect a 1-suboptimal vertex b.

      end if
    end for
  until b
  if c **then**
    stop algorithm
  end if
  $k := 1$
  while $|d(u, x_k) - d(u, y_k)| > \epsilon$ **and** $\left|P_{e_{x_k y_k}}\right| < n + 2$ **do**
    let $w_k \in V$ be a vertex such that $E_F(x_k, w_k)/\pi(N) = \rho(x_k)$
    define the points λ_k and μ_k by
      $d(u, \lambda_k) := d(u, x_k) + (F_{p-k-1}/F_{p-k+1}) d(x_k, y_k)$
      $d(u, \mu_k) := d(u, x_k) + (F_{p-k}/F_{p-k+1}) d(x_k, y_k)$
    $\rho_k := \min\{\rho(x_k), \rho_{k-1}\}$
    if $\rho(x_k) < \rho_{k-1}$ **then**
      $P_{e_{x_k y_k}} := \{x_k\}$ (with respect to the direction of counting)
      $P_e := \emptyset \quad \forall e \in E \setminus \{e_{x_k y_k}\}$
    else if $\rho(x_k) = \rho_{k-1}$ **then**
      $P_{e_{x_k y_k}} := P_{e_{x_k y_k}} \cup \{x_k\}$ (with respect to the direction of counting)
    end if
    if $E_F(\lambda_k, w_k)/\pi(N) < \rho_k > E_F(\mu_k, w_k)/\pi(N)$ **then**
      $E_{cand} := E_{cand} \cup \{\lambda_k y_k\}$ (with respect to the direction of counting)
      $x_{k+1} := x_k,\ y_{k+1} := \lambda_k$
    else if $E_F(\lambda_k, w_k)/\pi(N) \geq \rho_k > E(\mu_k, w_k)/\pi(N)$ **then**
      $x_{k+1} := \lambda_k,\ y_{k+1} := y_k$
    else if $E_F(\mu_k, w_k)/\pi(N) \geq \rho_k > E_F(y_k, w_k)/\pi(N)$ **then**
      $x_{k+1} := \mu_k,\ y_{k+1} := y_k$
    else if $E_F(\mu_k, w_k)/\pi(N) \geq \rho_k \leq E_F(y_k, w_k)/\pi(N)$ **then**
      $x_{k+1} := y_k,\ y_{k+1} := y_k$
    end if
    if $|d(u, x_{k+1}) - d(u, y_{k+1})| > \epsilon$ **then**
      exchange x_{k+1} and y_{k+1} as well as the direction of counting
      $k := k + 1$

else if $x_k \neq y_k$ **then**
  $\rho_0 := \rho(\alpha_k)$ where $d(u, \alpha_k) := \frac{1}{2}(d(u, x_k) + d(u, y_k))$
  if $\rho_0 < \rho_k$ **then**
    $P_{e_{x_k y_k}} := \{\alpha_k\}$ (with respect to the direction of counting)
    $P_e := \emptyset \quad \forall e \in E \backslash e_{x_k y_k}$
  else if $\rho_0 = \rho_k$ **then**
    $P_{e_{x_k y_k}} := P_{e_{x_k y_k}} \cup \{\alpha_k\}$ (with respect to the direction of counting)
  end if
else if $x_k = y_k$ **then**
  $\rho_0 := \rho_k$
  if $\rho(y_k) = \rho_k$ **then**
    $P_{e_{x_k y_k}} := P_{e_{x_k y_k}} \cup \{y_k\}$ (with respect to the direction of counting)
  end if
end if
end while
if $|d(u, x_k) - d(u, y_k)| > \epsilon$ **then**
  $E_{cand} := E_{cand} \cup \{x_k y_k\}$ (with respect to the direction of counting)
  $\rho_0 := \rho_k$
  $P_{e_{x_k y_k}} := \emptyset$
end if
end while

3.3 Including Edge Demand

In this section we extend the choice rule of Section 3.2 to include the concept of edge demand (see Chapter 1.3) as this may be a fruitful source for future research. In Section 3.3.1 we will show that the discretization properties of Section 3.2 do not readily carry over to the case of vertex and edge demand. Furthermore, we will see that, even for the simple class of chain networks with edge demand only, 1-optimal points can (in general) not be determined by straight forward analytical calculations (Section 3.3.2).

Proposition 3.3.1. *Let N be a (finite) network and let $[u,v]$ be any edge of the network and $x, y \in [0, \lambda(uv)]$ be any two points on this edge. Then*

$$\hat{e}^k_{[u,v]}(x_{uv}) := \frac{d(x_{uv}, y)^k}{d(x_{uv}, y)^k + d(x_{uv}, x)^k}$$

can be extended continuously to the interval $I = [0, \lambda(uv)]$ for all $k \in \mathbb{N} \cup \{\infty\}$ by defining

$$e^k_{[u,v]}(x_{uv}) := \begin{cases} \frac{1}{2} & \text{if } x_{uv} = x = y, \\ \frac{d(x_{uv},y)^k}{d(x_{uv},y)^k + d(x_{uv},x)^k} & \text{else.} \end{cases}$$

Proof. Note that

$$\hat{e}^k_{[u,v]}(x_{uv}) := \frac{d(x_{uv}, y)^k}{d(x_{uv}, y)^k + d(x_{uv}, x)^k}, \; x \neq y$$

is continuous in x_{uv} on I, since the functions $d(x_{uv}, y)$ and $d(x_{uv}, x)$ are continuous (cf. Section 3.1) and $d(x_{uv}, y)^k + d(x_{uv}, x)^k \neq 0$.

For $x = y \in (0, \lambda(uv))$ we get

$$\lim_{x_{uv} \searrow x} \hat{e}^k_{[u,v]}(x_{uv}) = \lim_{x_{uv} \searrow x} \frac{(x_{uv} - y)^k}{(x_{uv} - y)^k + (x_{uv} - x)^k} = \lim_{x_{uv} \searrow x} \frac{(x_{uv} - x)^k}{2(x_{uv} - x)^k} = \frac{1}{2},$$

$$\lim_{x_{uv} \nearrow x} \hat{e}^k_{[u,v]}(x_{uv}) = \lim_{x_{uv} \nearrow x} \frac{(y - x_{uv})^k}{(y - x_{uv})^k + (x - x_{uv})^k} = \lim_{x_{uv} \nearrow x} \frac{(x - x_{uv})^k}{2(x - x_{uv})^k} = \frac{1}{2}$$

(cf. Section 3.1). For $x = y = 0$ (and $x = y = \lambda(uv)$) only the first (second) limit exists. □

Thus, in analogy to Section 3.2, we may define

$$\bar{E}^k_L(x, y) := \begin{cases} \frac{1}{2} \cdot \delta(E) & \text{if } x = y, \\ \sum\limits_{[u,v] \in E} \int\limits_0^{\lambda(uv)} \delta(uv) \cdot \frac{d(x_{uv},y)^k}{d(x_{uv},y)^k + d(x_{uv},x)^k} \, dx_{uv} & \text{else,} \end{cases} \tag{3.8}$$

to be the expected number of edge customers visiting the leader's facility x for any given follower's location $y \in N$ and for all $k \in \mathbb{N} \cup \{\infty\}$. Analogously,

$$\bar{E}_F^k(x,y) := \begin{cases} \frac{1}{2} \cdot \delta(E) & \text{if } x = y, \\ \sum\limits_{[u,v] \in E} \int\limits_0^{\lambda(uv)} \delta(uv) \cdot \frac{d(x_{uv},x)^k}{d(x_{uv},y)^k + d(x_{uv},x)^k} \, dx_{uv} & \text{else,} \end{cases} \tag{3.9}$$

corresponds to the expected number of edge customers visiting the follower's facility. Hence, in line with our assumption in the context of vertex customers, we assume that, if $x = y$, ties are broken such that all edge customers (are expected to) accommodate half of their demand at each of the facilities.

Now, given a leader's facility $x \in N$ and a follower's location $y \in N$, we define

$$\hat{E}_q^k(x,y) := E_q^k(x,y) + \bar{E}_q^k(x,y) \tag{3.10}$$

to be the expected total number of (vertex and edge) customers visiting the facility of player $q \in \{L, F\}$ for any $k \in \mathbb{N} \cup \{\infty\}$.

We find

$$\lim_{k \to \infty} \frac{d(z,y)^k}{d(z,y)^k + d(z,x)^k} = \begin{cases} 1 & \text{if } d(z,x) < d(z,y), \\ 0.5 & \text{if } d(z,x) = d(z,y), \\ 0 & \text{if } d(z,x) > d(z,y), \end{cases}$$

for any triple $x, y, z \in N$. Hence, as before, the decisions of the customers become deterministic for very large k. Moreover, it is easy to see that Proposition 3.2.1 holds for $\hat{E}_q^k(x,y)$, $q \in \{L, F\}$, as well. Hence, Definitions 3.2.1 and 3.2.2 can easily be modified to adapt to the case of vertex and edge demand. The maximal relative k-rejection of a point $x \in N$ becomes

$$\hat{\rho}^k(x) := \max_{y \in N} \frac{\hat{E}_F^k(x,y)}{\xi(N)}. \tag{3.11}$$

A point $x \in N$ is k-optimal, if $\hat{E}_L^k(x,y) \geq \frac{1}{2}\xi(N)$ for all $y \in N$. A point x is k-suboptimal, if $\hat{\rho}^k(x) = \min\limits_{z \in N} \hat{\rho}^k(z)$.

3.3.1 Basic Properties

An obvious question that arises is as follows: Is the maximal relative 1-rejection of a point still reached at a vertex as it is the case when we consider vertex demand only? To answer this question, we consider an example network with edge demand only and $n = 2$, $\delta([1,2]) = 1$, $\lambda([1,2]) = 1$. We let global variables $x, y \in [0,1]$ define the players' locations in this simple chain network. Then it is easy to show that

$$\hat{E}_L(x,y) = \begin{cases} \frac{1}{2} & \text{if } x = y, \\ \frac{1}{2} + \frac{1}{4}(x-y) \cdot \ln\left(\frac{2-x-y}{y+x}\right) & \text{else.} \end{cases}$$

Now assume the leader to be located at $x' = 0$. Then $\hat{E}_L(x',0) = \hat{E}_L(x',1) = 0.5$, while $\hat{E}_L(x',y) < 0.5$ for all $y \neq 0$ and $y \neq 1$. Therefore, we can immediately conclude that the maximal relative 1-rejection of a point x may, in general, be only reached at points that do not correspond to vertices of the network.

While it is similarly easy to see that 1-optimal points need not necessarily be vertices when we consider networks with vertex and edge demand, one may wonder if 1-optimal points always exist on networks with edge demand only. Consider the following example with edge demand only: Let $k = 1$, $n = 4$, $\delta([1,2]) = \delta([3,4]) = 3$, $\delta([2,3]) = 2$ and $\lambda([1,2]) = \lambda([2,3]) = \lambda([3,4]) = 1$. Again, we consider global variables $x, y \in [0,3]$ defining the players' locations in this chain network. Then one can easily calculate $\hat{E}_L(x,y)$ for specific values of x and y. Figure 3.10 illustrates resulting regions in the xy-plane.

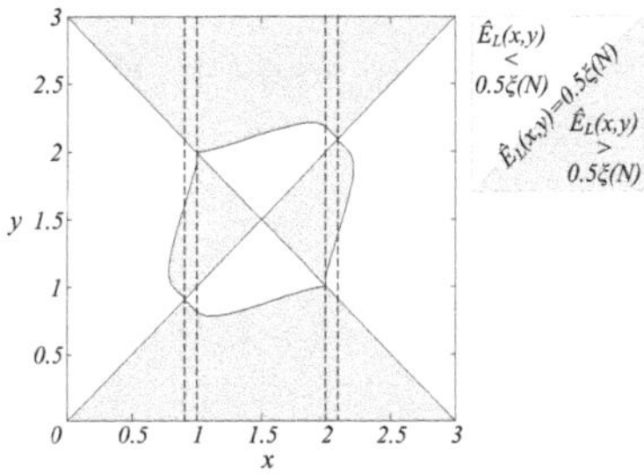

Fig. 3.10: No 1-optimal point

$\hat{E}_L(x,y) > \hat{E}_F(x,y)$ (gray), $\hat{E}_L(x,y) < \hat{E}_F(x,y)$ (white), and $\hat{E}_L(x,y) = \hat{E}_F(x,y)$ (border lines of the former two kinds of regions). Given a leader's location x', the follower will always be able to locate at $y' \neq x'$, such that $\hat{E}_L(x',y') < 0.5 \cdot \xi(N)$. As a consequence we may conclude that, given an arbitrary (finite) network N with edge demand only, 1-optimal points don't always exist.

3.3.2 1-Optimal Points on Chain Networks with Edge Demand Only

Let us restrict ourselves to the simple case of chain networks $N = (V, E, \lambda)$ with vertex set $V = \{u_1, ..., u_n\}$ and edge set $E = \{[u_i, u_{i+1}] | i = 1, ..., n-1\}$, assuming that there is no vertex demand, i.e. $\pi(V) = 0$. We use global variables $x, y \in [0, d(1,n)]$ to allow the definition of any leader's or follower's location on the chain network (see Figure 3.11). For the sake of notational convenience we will denote the vertices by natural numbers, i.e. i instead of u_i, whenever possible, because the distinction of vertex numbers and distinct values of the global variables will, in general, become clear from the context.

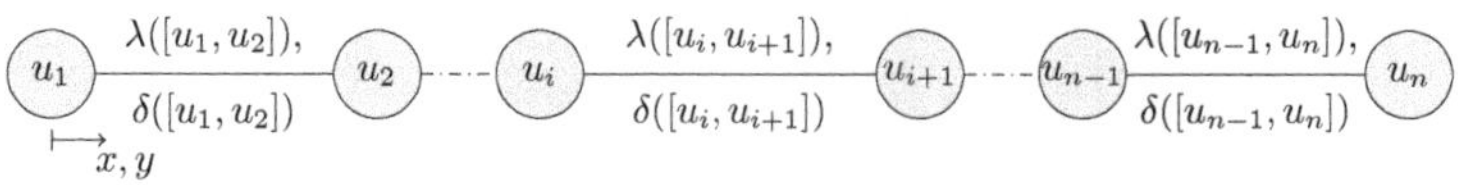

Fig. 3.11: Chain network

Let $x, y \in [d(1,u), d(1,v)]$. That is, x and y are two arbitrary points on edge $[u,v] \in E$. Then one determines

$$\hat{E}_L(x,y)|_{x,y\, on\, [u,v]} = \begin{cases} 0.5\xi(N) & \text{if } x = y \in \{d(1,u), d(1,v)\}, \\ \left.\begin{array}{l} \sum\limits_{i=1, i \neq u}^{n-1} \delta(i,i+1) \cdot (0.5\lambda(i,i+1) \\ +0.25(x-y) \cdot \ln\left(\frac{-2d(1,i+1)+x+y}{-2d(1,i)+x+y}\right)) \\ +\delta(uv) \cdot (0.5\lambda(uv) + 0.25(x-y) \\ \cdot \ln\left(\frac{2d(1,v)-x-y}{-2d(1,u)+x+y}\right)) \end{array}\right\} & \text{else.} \end{cases}$$

As we consider networks without vertex demand, we may restrict our attention to points $x = y$. This is eventually because the follower will always be able to co-locate so that he is expected to gain at least one half of the accumulated edge demand of the network. One easily derives

$$\left.\frac{\partial \hat{E}_L(x,y)}{\partial y}\right|_{x=y \text{ on } [u,v]} = \begin{cases} +\infty & \left.\begin{array}{c} \text{if } \delta(uv) > \delta(v,v+1) \\ \text{and } x = d(1,v) \\ \text{or} \\ \text{if } \delta(uv) < \delta(u-1,u) \\ \text{and } x = d(1,u), \end{array}\right\} \\[2ex] -\infty & \left.\begin{array}{c} \text{if } \delta(uv) < \delta(v,v+1) \\ \text{and } x = d(1,v) \\ \text{or} \\ \text{if } \delta(uv) > \delta(u-1,u) \\ \text{and } x = d(1,u), \end{array}\right\} \\[2ex] \begin{array}{l} \sum_{i=1}^{u-1} -\frac{1}{4}\delta(i,i+1) \cdot \ln\left(\frac{d(i+1,v)}{d(i,v)}\right) \\ + \sum_{i=v+1}^{n-1} -\frac{1}{4}\delta(i,i+1) \cdot \ln\left(\frac{d(v,i+1)}{d(v,i)}\right) \\ -\frac{1}{4}\delta(uv) \cdot \ln\left(\frac{\lambda(v,v+1)}{\lambda(uv)}\right) \end{array} & \left.\begin{array}{c} \text{if } \delta(uv) = \delta(v,v+1) \\ \text{and } x = d(1,v), \end{array}\right\} \\[2ex] \begin{array}{l} \sum_{i=1}^{u-2} -\frac{1}{4}\delta(i,i+1) \cdot \ln\left(\frac{d(i+1,u)}{d(i,u)}\right) \\ + \sum_{i=v}^{n-1} -\frac{1}{4}\delta(i,i+1) \cdot \ln\left(\frac{d(u,i+1)}{d(u,i)}\right) \\ -\frac{1}{4}\delta(uv) \cdot \ln\left(\frac{\lambda(uv)}{\lambda(u-1,u)}\right) \end{array} & \left.\begin{array}{c} \text{if } \delta(uv) = \delta(u-1,u) \\ \text{and } x = d(1,u), \end{array}\right\} \\[2ex] \begin{array}{l} \sum_{i=1, i\neq u}^{n-1} -\frac{1}{4}\delta(i,i+1) \cdot \ln\left(\frac{-d(1,i+1)+x}{-d(1,i)+x}\right) \\ -\frac{1}{4}\delta(uv) \cdot \ln\left(\frac{d(1,v)-x}{-d(1,u)+x}\right) \end{array} & \Big\} \text{ else,} \end{cases}$$

after defining $\delta(n,n+1) := 0$, $\delta(0,1) := 0$, $\lambda(0,1) := 1$ and $\lambda(n,n+1) := 1$. A necessary condition for a point x on edge $[u,v]$ to be 1-optimal is

$$\left.\frac{\partial \hat{E}_L(x,y)}{\partial y}\right|_{x=y \text{ on } [u,v]} = 0. \tag{3.12}$$

It follows that, when restricting our attention to the 1-optimality of vertices $u \in V$, we will only need to check first order partial derivatives if $\delta(u-1,u) = \delta(u,u+1)$. For all points x on edges $[u,v] \in E$, $x \notin V$ we get

$$\left(\frac{d(1,v)-x}{-d(1,u)+x}\right)^{\delta(uv)} \cdot \prod_{i=1, i\neq u}^{n-1} \left(\frac{-d(1,i+1)+x}{-d(1,i)+x}\right)^{\delta(i,i+1)} = 1 \qquad (3.13)$$

from (3.12). Finding the solutions to (3.13), however, will, in general, require the use of numerical methods. Additionally, after having found all solutions to (3.13) for all edges $[u,v] \in E$, we will have to check the second order partial derivatives, as the follower will always minimize the leader's expected gain. We are then left with a set of potentially 1-optimal locations X_{pot} and the task of proving or disproving their global optimality with respect to the functions $\hat{E}_L(\hat{x},y)$ for $\hat{x} \in X_{pot}$. Unfortunately, there does not seem to be an exploitable special structure to these nonlinear functions. Hence, once more, we are dependent on the performance of general numerical methods.

3.4 Conclusion

This chapter has been concerned with voting location problems under a proportional choice rule that has previously been introduced by Bauer et al. (1993) with vertex demand only. We have refined a known discretization property by proving convexity and concavity properties of related expected payoff functions. Moreover, we have answered the long time open question whether 1-suboptimal points are always vertices by providing a counterexample on a tree network. However, there may well exist a finite dominating set that future research might explore.

A discretization property (vertex optimality property) for the family of connected networks with two vertices only has been derived. This property immediately raises an interesting question for future research: Does the vertex optimality property carry over to the more complex family of chain networks $N = (V, E, \lambda)$ with vertex set $V = \{u_1, ..., u_n\}$ and edge set $E = \{[u_i, u_{i+1}] | i = 1, ..., n-1\}$ for $k = 1$? As before, we use a global variable

x to define any point of the chain network (Figure 3.11). Furthermore, for the sake of notational convenience, we denote the vertices by natural numbers, i.e. i instead of u_i. Then, a first related result is as follows.

Proposition 3.4.1. *Let $N = (V, E, \lambda)$ be a chain network as defined above. Add an additional vertex $n+1$ with $\pi(n+1) > 0$ and connect it to vertex n via edge $[n, n+1]$ with $\lambda([n, n+1]) > 0$. Denote the augmented chain network by $N' = (V', E', \lambda')$ and define $\Delta E_F(x, w) := E'_F(x, w) - E_F(x, w)$ for any $w \in V$ and for $x \leq d(1, n)$, where E'_F denotes the follower's (expected) market share in the augmented chain network. Then*

1. $\Delta E_F(x, w) > 0$,
2. $\Delta E_F(x_1, w) - \Delta E_F(x_2, w) > 0$ *for* $x_1 < x_2$,
3. $\Delta E_F(x, w_1) - \Delta E_F(x, w_2) > 0$ *for* $w_1, w_2 \in V$, $d(1, w_1) > d(1, w_2)$.

Proof. It is easy to see that

$$\Delta E_F(x, w) = \pi(n+1) \cdot \frac{d(1, n+1) - x}{d(1, n+1) - x + d(w, n+1)} > 0$$

and

$$\frac{d(\Delta E_F(x, w))}{dx} = \pi(n+1) \cdot \frac{-d(w, n+1)}{(d(1, n+1) - x + d(w, n+1))^2} < 0.$$

The claims follow immediately. □

In the second part of this chapter we have been concerned with implementational issues of an algorithm for determining 1-suboptimal points that has been proposed by Bauer et al. (1993). Moreover, we have shown that the proportional choice rule can easily be extended to include edge demand. We have seen that the discretization results for the case of vertex demand only do not readily carry over to this more general case, so that there remain interesting questions to be studied in future.

Chapter 4

$(r|p)$-Centroid Problems on Networks with Vertex and Edge Demand

As mentioned in Chapters 1 and 2, Dasci et al. (2002) and Okunuki and Okabe (2002) were among the first to consider edge demand in competitive location problems on general networks. While these authors consider $(r|X_p)$-medianoid problems, it is the aim of this chapter to apply the concept of edge demand to $(r|p)$-centroid problems, with a focus on drawing the line between hard and easy problem classes as in Kress and Pesch (2012c).

We consider a binary choice rule throughout this chapter. That is, we assume that each customer chooses the closest facility to accommodate all of his demand. We additionally assume that ties are broken in favor of the leader. As a direct consequence we may assume that $p + r \leq n$ for the discrete version of the centroid problem under consideration. When referring to a discrete problem we assume the potential facility sites to be restricted to the vertex set of the network.

In Section 4.1 we analyze the computational complexity of the binary $(r|p)$-centroid problem on general networks with vertex and edge demand and present some basic properties as well as mathematical programming formulations for the discrete problem class. An efficient algorithm for the discrete, binary $(1|p)$-centroid problem on tree networks with vertex and edge demand is given in Section 4.2. The chapter closes with a conclusion in Section 4.3.

4.1 The Binary (r|p)-Centroid Problem with Vertex and Edge Demand

This section is structured as follows. Section 4.1.1 presents bilevel programming formulations for the discrete, binary $(r|p)$-centroid problem with vertex and edge demand. Some computational results for the corresponding follower's problem are given in Section 4.1.2 in order to illustrate the need for approximate solution procedures when designing heuristic algorithms for the discrete leader's problem. In Section 4.1.3 we derive complexity results for the discrete and continuous versions of the binary $(1|p)$-centroid problem with edge demand only. Section 4.1.4 is concerned with the identification of finite dominating sets.

4.1.1 Bilevel Programming Models

Consider the discrete, binary $(r|p)$-centroid problem with vertex and edge demand and assume – without loss of generality – that the underlying network is complete. If this is not the case, we simply add missing edges with infinite edge lengths and zero demand densities.

We define the following variables:

$$x_{ij}^F := \begin{cases} 1 & \text{if the vertex customers in vertex i are} \\ & \text{served by a follower's facility in vertex j,} \\ 0 & \text{else,} \end{cases} \qquad \forall\, i,j \in V, \quad (4.1)$$

$$x_{ij}^L := \begin{cases} 1 & \text{if the vertex customers in vertex i are} \\ & \text{served by a leader's facility in vertex j,} \\ 0 & \text{else,} \end{cases} \qquad \forall\, i,j \in V, \quad (4.2)$$

$$y_j^F := \begin{cases} 1 & \text{if the follower locates in vertex j,} \\ 0 & \text{else,} \end{cases} \qquad \forall\, j \in V, \quad (4.3)$$

$$y_j^L := \begin{cases} 1 & \text{if the leader locates in vertex j,} \\ 0 & \text{else,} \end{cases} \qquad \forall j \in V. \tag{4.4}$$

For all $i, j = 1, ..., n$, let C_{ij} and $\bar{C}_{ij}$ be the sets of vertices k with $d(k,i) < d(i,j)$ or $d(k,i) = d(i,j)$, respectively (cf. also Dobson and Karmarkar, 1987). Furthermore, define $\hat{C}_{ij} := C_{ij} \cup \bar{C}_{ij}$. We can now state a binary bilevel nonlinear programming model (BBNP) for the problem under consideration.

$$\max_{\mathbf{x^L},\mathbf{y^L}} \quad \sum_{i=1}^{n}\sum_{j=1}^{n} x_{ij}^L \left[\pi(i) + \sum_{k=1}^{n} \tfrac{1}{2}\delta(i,k)\left(\lambda(i,k) - d(i,j) + \sum_{l=1}^{n}(x_{kl}^L + x_{kl}^F)d(k,l)\right)\right] \tag{4.5}$$

$$\text{subject to} \quad \sum_{j=1}^{n} y_j^L = p, \tag{4.6}$$

$$x_{ij}^L \leq y_j^L \qquad \forall\, i,j = 1,...,n, \tag{4.7}$$

$$\sum_{j=1}^{n} x_{ij}^L \leq 1 \qquad \forall\, i = 1,...,n, \tag{4.8}$$

$$x_{ij}^L + y_k^L \leq 1 \qquad \forall\, i,j = 1,...,n,\ k \in C_{ij}, \tag{4.9}$$

$$x_{ij}^L + y_k^F \leq 1 \qquad \forall\, i,j = 1,...,n,\ k \in C_{ij}, \tag{4.10}$$

and $\mathbf{x^F}$, $\mathbf{y^F}$ being a solution of

$$\max_{\mathbf{x^F},\mathbf{y^F}} \quad \sum_{i=1}^{n}\sum_{j=1}^{n} x_{ij}^F \left[\pi(i) + \sum_{k=1}^{n} \tfrac{1}{2}\delta(i,k)\left(\lambda(i,k) - d(i,j)\sum_{l=1}^{n} + \sum_{l=1}^{n}(x_{kl}^L + x_{kl}^F)d(k,l)\right)\right] \tag{4.11}$$

$$\text{subject to} \quad \sum_{j=1}^{n} y_j^F = r, \tag{4.12}$$

$$x_{ij}^F \leq y_j^F \qquad \forall\, i,j = 1,...,n, \tag{4.13}$$

$$\sum_{j=1}^{n} \left(x_{ij}^F + x_{ij}^L\right) = 1 \qquad \forall\, i = 1,...,n, \tag{4.14}$$

$$x_{ij}^F + y_k^L \leq 1 \qquad \forall\, i,j = 1,...,n,\ k \in \hat{C}_{ij}, \tag{4.15}$$

$$x_{ij}^F + y_k^F \leq 1 \qquad \forall\, i,j = 1,...,n,\ k \in C_{ij}, \tag{4.16}$$

$$x_{ij}^L, x_{ij}^F \in \{0,1\} \qquad \forall\, i,j = 1,...,n, \tag{4.17}$$

$$y_j^L, y_j^F \in \{0,1\} \qquad \forall\, j = 1,...,n. \tag{4.18}$$

The objective functions (4.5) and (4.11) correspond to the maximization of the market share of each of the players, taking into account vertex and edge demand. Constraints (4.6) and (4.12) ensure that the players locate exactly p and r facilities, respectively. Conditions (4.7) and (4.13) guarantee that a (vertex) customer can only be served if the corresponding facility has been opened. Constraints (4.8) and (4.14) relate to the fact that the demand of each (vertex) customer must be served by exactly one facility, while conditions (4.9), (4.10), (4.15) and (4.16) guarantee that each (vertex) customer is served by its closest facility (cf. Dobson and Karmarkar, 1987). Observe that edge customers are guaranteed to be served implicitly. Figure 4.1 depicts an example. Vertex i is served by a leader's facility in vertex j, while vertex k is served by a follower's facility in vertex l. Therefore, there must exist a point $x_{ik} = x$ on edge $[i, k]$, where $d(i, j) + x = \lambda(i, k) - x + d(k, l)$. Every edge customer located at $x_{ik} \leq x$ ($x_{ik} > x$) will be served by the leader (follower).

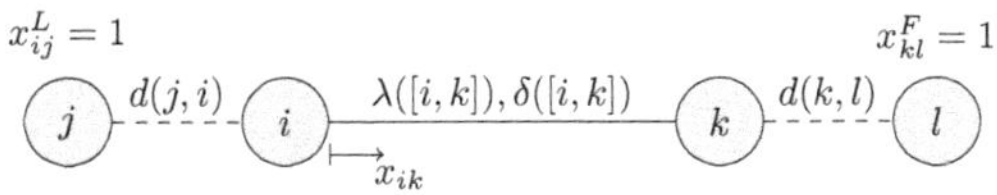

Fig. 4.1: Serving of edge customers

Observe that

$$\sum_{i=1}^{n}\sum_{j=1}^{n} x^L_{ij}\left[\pi(i) + \sum_{k=1}^{n}\frac{1}{2}\delta(i,k)\left(\lambda(i,k) - d(i,j) + \sum_{l=1}^{n}(x^L_{kl} + x^F_{kl})d(k,l)\right)\right]$$

$$= \sum_{i=1}^{n}\sum_{j=1}^{n} x^L_{ij}\pi(i) + \sum_{i=1}^{n}\sum_{j=1}^{n}\sum_{k=1}^{n}\frac{1}{2}\delta(i,k)x^L_{ij}\left(\lambda(i,k) - d(i,j)\right)$$

$$+ \sum_{i=1}^{n}\sum_{j=1}^{n}\sum_{k=1}^{n}\sum_{l=1}^{n}\frac{1}{2}\delta(i,k)x^L_{ij}x^L_{kl}d(k,l)$$

$$+ \sum_{i=1}^{n}\sum_{j=1}^{n}\sum_{k=1}^{n}\sum_{l=1}^{n}\frac{1}{2}\delta(i,k)x^L_{ij}x^F_{kl}d(k,l).$$

An analogous result holds for the follower's objective function. Thus, we may linearize BBNP by defining binary variables z^{LL}_{ijkl}, z^{LF}_{ijkl}, z^{FL}_{ijkl} and z^{FF}_{ijkl} for all $i, j, k, l = 1, ..., n$ and introducing additional constraints. This results in a binary bilevel linear programming model (BBLP).

$$\begin{aligned}\max_{\mathbf{x^L},\mathbf{y^L}} \quad & \sum_{i=1}^{n}\sum_{j=1}^{n} x^L_{ij}\pi(i) \\ & +\sum_{i=1}^{n}\sum_{j=1}^{n}\sum_{k=1}^{n} \tfrac{1}{2}\delta(i,k)x^L_{ij}\left(\lambda(i,k)-d(i,j)\right) \\ & +\sum_{i=1}^{n}\sum_{j=1}^{n}\sum_{k=1}^{n}\sum_{l=1}^{n} \tfrac{1}{2}\delta(i,k)z^{LL}_{ijkl}d(k,l) \\ & +\sum_{i=1}^{n}\sum_{j=1}^{n}\sum_{k=1}^{n}\sum_{l=1}^{n} \tfrac{1}{2}\delta(i,k)z^{LF}_{ijkl}d(k,l)\end{aligned} \tag{4.19}$$

subject to constraints (4.6)-(4.10),

$$z^{LL}_{ijkl} \leq \tfrac{1}{2}(x^L_{ij} + x^L_{kl}) \qquad \forall\, i,j,k,l = 1, ..., n, \tag{4.20}$$

$$z^{LF}_{ijkl} \leq \tfrac{1}{2}(x^L_{ij} + x^F_{kl}) \qquad \forall\, i,j,k,l = 1, ..., n, \tag{4.21}$$

and $\mathbf{x^F}$, $\mathbf{y^F}$ being a solution of

$$\begin{aligned}\max_{\mathbf{x^F},\mathbf{y^F}} \quad & \sum_{i=1}^{n}\sum_{j=1}^{n} x^F_{ij}\pi(i) \\ & +\sum_{i=1}^{n}\sum_{j=1}^{n}\sum_{k=1}^{n} \tfrac{1}{2}\delta(i,k)x^F_{ij}\left(\lambda(i,k)-d(i,j)\right) \\ & +\sum_{i=1}^{n}\sum_{j=1}^{n}\sum_{k=1}^{n}\sum_{l=1}^{n} \tfrac{1}{2}\delta(i,k)z^{FL}_{ijkl}d(k,l) \\ & +\sum_{i=1}^{n}\sum_{j=1}^{n}\sum_{k=1}^{n}\sum_{l=1}^{n} \tfrac{1}{2}\delta(i,k)z^{FF}_{ijkl}d(k,l)\end{aligned} \tag{4.22}$$

subject to constraints (4.12)-(4.18),

$$z^{FL}_{ijkl} \leq \tfrac{1}{2}(x^F_{ij} + x^L_{kl}) \qquad \forall\, i,j,k,l = 1, ..., n, \tag{4.23}$$

$$z^{FF}_{ijkl} \leq \tfrac{1}{2}(x^F_{ij} + x^F_{kl}) \qquad \forall\, i,j,k,l = 1, ..., n, \tag{4.24}$$

$$z^{LL}_{ijkl},\ z^{LF}_{ijkl} \in \{0,1\} \qquad \forall\, i,j,k,l = 1, ..., n, \tag{4.25}$$

$$z^{FL}_{ijkl},\ z^{FF}_{ijkl} \in \{0,1\} \qquad \forall\, i,j,k,l = 1, ..., n. \tag{4.26}$$

Constraints (4.20), (4.21), (4.23) and (4.24) can easily be rearranged to receive a disaggregated model (BBLP2). Inequalities (4.20), for example, may be replaced by

$$z_{ijkl}^{LL} \leq x_{ij}^{L} \quad \forall\, i,j,k,l = 1,...,n, \tag{4.27}$$

$$z_{ijkl}^{LL} \leq x_{kl}^{L} \quad \forall\, i,j,k,l = 1,...,n. \tag{4.28}$$

4.1.2 Potential Solution Approaches

Due to their hierarchical structure, bilevel programming problems are usually very hard to solve (cf. Chapter 2.6). Indeed, as we will see in the following section, the discrete, binary $(1|p)$-centroid problem on a general network is NP-hard if we consider vertex demand only, edge demand only or vertex and edge demand. Thus, the design of exact solution approaches does not seem to be very promising.[1] Heuristic solution approaches will need to compute or approximate the follower's response to some given set of leader's locations. Since the discrete, binary $(r|X_1)$-medianoid problem is known to be NP-hard on general networks (again, for the case of vertex demand only, edge demand only or vertex and edge demand, cf. Chapter 2.3), calculating the follower's response is most likely too expensive with regard to computational time. To support this assumption, we have coded the model BBNP in C++ for fixed sets of leader's locations. The resulting quadratic programs were solved with CPLEX 12.2 with a time limit of 1800 seconds. We generated a total of 50 test networks; ten complete networks for $n = 5, 10, 15, 20, 25$, respectively. Integer edge lengths, edge demand densities and vertex demands with $\lambda([u,v]) \in [1,5]$, $\delta([u,v]) \in [0,10]$ and $\pi(u) \in [0,10]$ for all $u, v \in V$ were generated randomly. Figures 4.2a and 4.2b plot the average computational time for the 5 network sizes with $r = 1,2,3,4$ and $p = 1$ or $p = 2$ random leader locations. Details are given in Table 4.1.[2] When generating or solving additional test instances with $n \geq 40$, we frequently ran out of memory.

1 See Roboredo and Pessoa (2013) and the references therein for exact solution approaches for the discrete, binary $(r|p)$-centroid problem with vertex demand only.

2 Entries "-" in the tables mark trivial instances with $r + p \geq n$. t_{avg} refers to the average computational time.

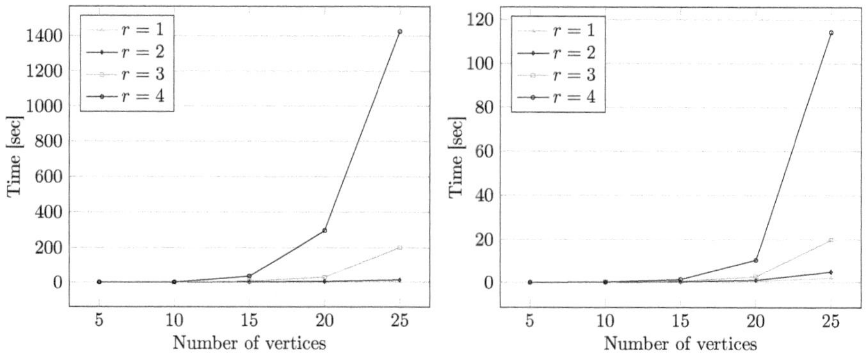

(a) $p = 1$ *(b) $p = 2$*

Fig. 4.2: Average comp. time for solving BBNP with fixed sets of leader's locations

Tab. 4.1: Computational results - solving BBNP with fixed sets of leader's locations

		$p = 1$			$p = 2$		
		t_{avg}	inst. not solved to opt.		t_{avg}	inst. not solved to opt.	
r	n	[sec.]	#	average gap [%]	[sec.]	#	average gap [%]
1	5	0.0188	0	0	0.0157	0	0
	10	0.1062	0	0	0.0875	0	0
	15	0.3963	0	0	0.2605	0	0
	20	1.0048	0	0	0.6396	0	0
	25	2.6973	0	0	2.0358	0	0
2	5	0.0967	0	0	0.0188	0	0
	10	0.4306	0	0	0.0828	0	0
	15	1.5272	0	0	0.3199	0	0
	20	3.9342	0	0	1.0142	0	0
	25	11.8092	0	0	4.7645	0	0
3	5	0.0361	0	0	-	-	-
	10	0.2915	0	0	0.1014	0	0
	15	6.0265	0	0	0.5302	0	0
	20	28.1928	0	0	2.672	0	0
	25	202.638	0	0	19.9147	0	0
4	5	-	-	-	-	-	-
	10	0.4084	0	0	0.1329	0	0
	15	35.2106	0	0	1.2654	0	0
	20	297.308	0	0	10.329	0	0
	25	1424.61	4	29.05	114.313	0	0

Since computational times are fairly large for test instances on relatively small networks, it seems to be reasonable to approximate the follower's response when designing heuristic algorithms for the discrete, binary $(r|p)$-centroid problem on networks with vertex and edge demand. Therefore, we refer to the tabu search algorithms by Benati and Laporte (1994) for discrete, binary $(r|p)$-centroid and $(r|X_p)$-medianoid problems on networks with vertex demand only. These algorithms can be adapted to our case (cf. Kress et al., 2012), since the objective function of the discrete, binary $(r|X_p)$-medianoid problem with vertex demand only remains submodular and non-decreasing when additionally incorporating edge demand.

Lemma 4.1.1. *The objective function of the discrete, binary $(r|X_p)$-medianoid problem with vertex and edge demand is submodular.*

Proof. Let Y_r be a solution to a discrete, binary $(r|X_p)$-medianoid problem with vertex and edge demand. Furthermore, let $S \subset T \subset Y_r$ and $k \in V \backslash T$. We want to show that $W_F(T \cup \{k\}|X_p) - W_F(T|X_p) \leq W_F(S \cup \{k\}|X_p) - W_F(S|X_p)$.

We define $\Delta_v(T,k)$ ($\Delta_v(S,k)$) to be the increase in the follower's market share that is induced by vertex customer $v \in V$ when augmenting T (S) with an additional facility in vertex k. Similarly, $\Delta_{[u,v]}(T,k)$ ($\Delta_{[u,v]}(S,k)$) is defined to be the increase in the follower's market share that is induced by the demand over edge $[u,v]$. Therefore, $W_F(T \cup \{k\}|X_p) - W_F(T|X_p) = \sum_{v \in V} \Delta_v(T,k) + \sum_{[u,v] \in E} \Delta_{[u,v]}(T,k)$ and $W_F(S \cup \{k\}|X_p) - W_F(S|X_p) = \sum_{v \in V} \Delta_v(S,k) + \sum_{[u,v] \in E} \Delta_{[u,v]}(S,k)$.

One of the following cases holds for every vertex $v \in V$:

1. $D(v,T) \geq D(v,X_p)$
   (a) and $d(v,k) \geq D(v,X_p)$,
   (b) and $d(v,k) < D(v,X_p)$,
2. $D(v,T) < D(v,X_p)$, $D(v,S) \geq D(v,X_p)$
   (a) and $d(v,k) \geq D(v,X_p)$,

   (b) and $d(v,k) < D(v,X_p)$, $d(v,k) \geq D(v,T)$,
   (c) and $d(v,k) < D(v,X_p)$, $d(v,k) < D(v,T)$,

3. $D(v,S) < D(v,X_p)$
   (a) and $d(v,k) \geq D(v,S)$,
   (b) and $d(v,k) < D(v,S)$, $D(v,S) = D(v,T)$,
   (c) and $d(v,k) < D(v,S)$, $D(v,S) > D(v,T)$.

Thus, for an arbitrary vertex $v \in V$, we have $\Delta_v(S,k) = \Delta_v(T,k) = 0$ for cases 1.a, 2.a, 3.a, 3.b and 3.c; $\Delta_v(S,k) = \Delta_v(T,k) = \pi(v)$ for case 1.b; $\Delta_v(S,k) = \pi(v) \geq \Delta_v(T,k) = 0$ for cases 2.b and 2.c. Similarly, for an arbitrary edge $[u,v] \in E$, we have $\Delta_{[u,v]}(T,k) \leq \Delta_{[u,v]}(S,k)$. The related case differentiation is fairly simple but rather lengthy (36 possible combinations of cases 1.a – 3.c). Hence, for the sake of brevity, we leave it to the reader. □

Obviously, if $S \subset T$, then $W_F(S|X_p) \leq W_F(T|X_p)$. This is a direct consequence of the non-negativity of the vertex demands and edge demand densities. Hence, analogously to Benati and Laporte (1994), one may apply Theorem 9.3 of Nemhauser and Wolsey (1999, Chap. III.3). Therefore, $W_F^*(r|X_p) \leq (1-[(r-1)/r]^r)^{-1} W_F^g(r|X_p)$, where $W_F^g(r|X_p)$ is the follower's market share obtained through a greedy heuristic.

Another promising approach might be the use of genetic algorithms, as the study by Jaramillo et al. (2002) indicates (cf. also Arostegui Jr. et al., 2006).

4.1.3 Some Complexity Results

This section aims at extending some well known complexity results on $(r|p)$-centroid problems (cf. already Chapter 2.4). From Hakimi (1983, 1990) and Spoerhase and Wirth (2009c), who considered vertex demand only, we can immediately conclude:

Theorem 4.1.1. *The problem of finding a continuous, binary* $(1|p)$*-centroid of a network* $N = (V, E, \lambda)$ *with vertex and edge demand is NP-hard.*

Proof. The NP-hard (Hakimi, 1983, 1990) continuous, binary $(1|p)$-centroid problem with vertex demand only is a special case of the continuous, binary $(1|p)$-centroid problem with vertex and edge demand. □

In what follows, we will show that the problem remains NP-hard even if we drop the vertex weights. This is in line with the results of Dasci et al. (2002) for the continuous, binary $(r|X_1)$-medianoid problem with (vertex and) edge demand (cf. Chapter 2.3).

Theorem 4.1.2. *The problem of finding a continuous, binary* $(1|p)$*-centroid of a network* $N = (V, E, \lambda)$ *with edge demand only, i.e.* $\pi(v) = 0$ *for all* $v \in V$*, is NP-hard.*

Proof. The proof is in analogy to Hakimi (1990) who considers networks with vertex demand only. The theorem is proven by reducing the *Vertex Cover* (VC) problem (cf. Garey and Johnson, 1979, p. 190).

Vertex Cover Problem: Given a graph $G = (V, E)$ and an integer $p \leq |V|$, is there a subset $V' \subseteq V$ with $|V'| \leq p$ such that for each edge $e = [u, v] \in E$ at least one of u and v belongs to V'?

Consider an instance $I(VC)$ of the VC problem on graph $G = (V, E)$ and replace each edge $e_j = [u, v] \in E$ by the structure shown in Figure 4.3 ("diamond structure joining u and v", Hakimi, 1990) to construct a network $N_1 = (V_1, E_1, \lambda)$ with the corresponding edge lengths λ and edge demand densities δ.

We will show that there exists a set of p points X_p on N_1 such that $W_F(Y_1(X_p)|X_p) \leq 1$ for every point $Y_1(X_p)$ on N_1 if and only if the VC instance has a solution. This will prove the theorem.

"If": Suppose V' is a solution of $I(VC)$ and let $X_p = V'$ on N_1. Then for any diamond structure joining u and v in N_1, either u, v or both belong to $V' = X_p$. Thus, we obviously have $W_F(Y_1(X_p)|X_p) \leq 1$ for any point $Y_1(X_p)$ on N_1 (recall that ties are broken in favor of the leader).

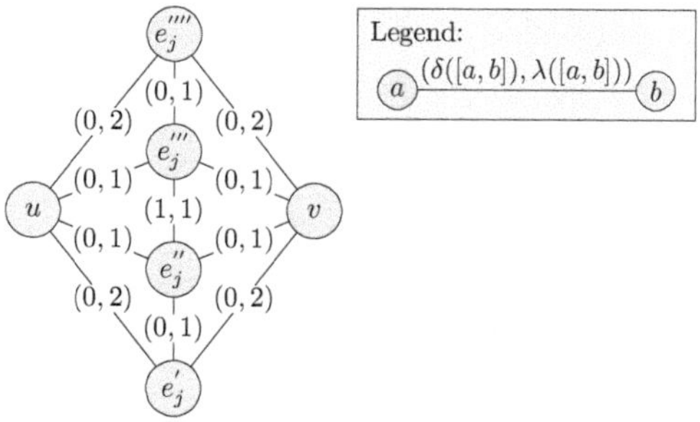

Fig. 4.3: Diamond structure

"Only if": Assume that a set of p points X_p on network N_1 is such that $W_F(Y_1(X_p)|X_p) \leq 1$ for every point $Y_1(X_p)$ on N_1. If there exists at least one point of X_p on each diamond of N_1, then one can move these points to ($\leq p$) vertices $V' \subseteq V$ such that each diamond has at least one vertex in V'. Then V' is a solution to $I(VC)$. Therefore, we may assume that there exists a diamond structure in N_1 joining, say, u and v such that this diamond structure contains no point of X_p. Suppose, $\min\{D(u,X_p),D(v,X_p)\} > 2$. It is immediately implied that $W_F(u|X_p) \geq 2$ or $W_F(v|X_p) \geq 2$ so that we may assume $0 < \min\{D(u,X_p),D(v,X_p)\} \leq 2$. Without loss of generality, let $\min\{D(u,X_p),D(v,X_p)\} = D(u,X_p)$. Let us say there exists another diamond in N_1 that contains a leader's facility defining $D(u,X_p)$ and corresponds to edge $\hat{e} = [u,v'] \in E$ and thus joins u and v', where $v \neq v'$. If there is exactly one point $x_p \in X_p$ on this diamond, then we can always select $Y_1(X_p)$ to lie on an edge of this diamond incident to u such that $W_F(Y_1(X_p)|X_p) > 1$:

1. If x_p is a point on edge $[u,\hat{e}'''']$ (or, analogously, on edge $[u,\hat{e}']$) with $0 < d(u,x_p) < 2$, then we select $Y_1(X_p) = u$.

2. If x_p is a point on edge $[u,\hat{e}''']$ (or, analogously, on edge $[u,\hat{e}'']$) with $0 < d(u,x_p) < 1$, then we choose $Y_1(X_p) = u$ as well.

3. If x_p is a point on edge $[v',\hat{e}''']$ or $[\hat{e}''',\hat{e}'''']$ (including the vertices v', $\hat{e}'''$ and $\hat{e}''''$), then we select $Y_1(X_p)$ to lie on edge $[u,\hat{e}'']$ such that $d(u,Y_1(X_p)) < D(u,X_p)$. The case where x_p is a point on edge $[v',\hat{e}'']$ or $[\hat{e}',\hat{e}'']$ is treated analogously (symmetry of the diamond).

4. If x_p is a point on edge $[\hat{e}'', \hat{e}''']$ with $x_p \neq \hat{e}''$ and $x_p \neq \hat{e}'''$, then we select $Y_1(X_p) = \hat{e}''$ or $Y_1(X_p) = \hat{e}'''$.

Therefore, there must exist at least two points of X_p on the diamond structure joining u and v'. This implies that, if there exist no points of X_p on an arbitrary diamond structure of N_1, then there are at least two points of X_p on one "adjacent" diamond structure of N_1. Therefore, there are enough points to be moved to all diamonds of N_1 so that the existence of a solution to VC is established. □

The hardness results that have been subject of this section carry over to the discrete versions of the related $(1|p)$-centroid problems. Spoerhase and Wirth (2009c) explicitly show this for the discrete, binary $(1|p)$-centroid problem with vertex demand only on a pathwidth bounded graph so that, analogously to Theorem 4.1.1, we can conclude:

Theorem 4.1.3. *The problem of finding a discrete, binary* $(1|p)$*-centroid of a network* $N = (V, E, \lambda)$ *with vertex and edge demand is NP-hard.*

Moreover, we note that the proof of Theorem 4.1.2 can easily be adapted for the discrete, binary $(1|p)$-centroid problem with edge demand only. Therefore, we have:

Theorem 4.1.4. *The problem of finding a discrete, binary* $(1|p)$*-centroid of a network* $N = (V, E, \lambda)$ *with edge demand only is NP-hard.*

Proof. We augment the diamond structure of Figure 4.3 by inserting vertices e_j^1, e_j^2, e_j^3 and e_j^4 at the midpoints of the edges $[u, e_j'']$, $[u, e_j''']$, $[v, e_j'']$ and $[v, e_j''']$.

The proof of Theorem 4.1.2 remains unchanged apart from the case differentiation: If there lies exactly one point $x_p \in X_p$ on the diamond joining u and v', then we can always select $Y_1(X_p)$ to lie on a vertex of this diamond such that $W_F(Y_1(X_p)|X_p) > 1$:

1. If $x_p = v'$, $x_p = \hat{e}^1$, $x_p = \hat{e}^2$, $x_p = \hat{e}^3$, $x_p = \hat{e}^4$, $x_p = \hat{e}'$ or $x_p = \hat{e}''''$ then we select $Y_1(X_p) = u$.
2. If $x_p = \hat{e}''$ then we select $Y_1(X_p) = \hat{e}^2$.

3. If $x_p = \hat{e}'''$ then we select $Y_1(X_p) = \hat{e}^1$.

The conclusion of the proof of Theorem 4.1.2 is therefore still true. □

Table 4.2 summarizes the complexity results.

Tab. 4.2: NP-hard binary $(1|p)$*-centroid problems*

type of demand	continuous	discrete
vertex demand only	NP-hard on general networks (Hakimi, 1990) NP-hard on pathwidth bounded graphs (Spoerhase and Wirth, 2009c)	NP-hard on general networks (Hakimi, 1990) NP-hard on pathwidth bounded graphs (Spoerhase and Wirth, 2009c)
vertex and edge demand	NP-hard on general networks and pathwidth bounded graphs (implied)	
edge demand only	NP-hard on general networks (Theorem 4.1.2)	NP-hard on general networks (Theorem 4.1.4)

Polynomial time algorithms are due to Hansen and Labbé (1988) and Spoerhase and Wirth (2009c) (cf. already Chapter 2.4). We extend these results in Section 4.2 by providing an efficient algorithm for the discrete, binary $(1|p)$-centroid problem on tree networks with vertex and edge demand.

4.1.4 Finite Dominating Sets

In Chapter 2 we have referred to discretization results for various sequential competitive location problems. An optimal solution to the continuous, binary $(r|1)$-centroid problem of a general network with vertex demand only, for example, is always a vertex for $r \geq 2$, while this is not the case for $r = 1$. If we restrict the network to be a tree, however, there always exists a vertex that is a $(1|1)$-centroid. It is easy to see that the former result carries over to the case of vertex and edge demand, while the latter does not, i.e. there exist tree networks with vertex and edge demand without a $(1|1)$-centroid located in a vertex (consider, for example, Figure 1.3 with $\pi(u) = \pi(v) = 0$ and $\lambda(uv) = \delta(uv) = 1$).

Proposition 4.1.1. *Let $r \geq 2$. Then there always exists a vertex $v \in V$ that is an optimal solution to the continuous, binary $(r|1)$-centroid problem with vertex and edge demand, defined on a network $N = (V, E, \lambda)$.*

The proof is straight forward. Assume that the leader locates in an arbitrary point $x \notin V$ of the network. Then there exists an (ϵ-) optimal solution to the corresponding medianoid problem with two follower's facilities located infinitesimally close to x such that the leader's gain is (arbitrarily close to) zero. Locating the leader's facility in a vertex $v \in V$, in contrast, will enforce the corresponding vertex customers to accommodate their demand at the leader's facility.

Spoerhase and Wirth (2009c) derive a finite dominating set for the continuous, binary $(1|p)$-centroid problem on tree networks with vertex demand only. They restrict $\lambda : E \to \mathbb{Q}^+$ and $\pi : V \to \mathbb{Q}_0^+$. Thus, they may assume without loss of generality that the edge lengths and vertex demands are integer numbers (cf. Section 4.2). The authors show that there always exists an optimal solution to the leader's problem, such that the distance from any of the leader's facilities to any vertex is an integer number or an integer number divided by two. Note that this discretization result is not polynomial. While, as described in Chapter 2.3, Dasci et al. (2002) are able to show that the discretization result by Megiddo et al. (1983) for the binary $(r|X_p)$-medianoid problem with vertex demand only almost directly applies when additionally considering edge demand, the same is implausible for the finite dominating set just described for the continuous, binary $(1|p)$-centroid problem on tree networks. Consider, for instance, a chain network $N = (V, E, \lambda)$ with $V = \{1, 2, 3\}$, $E = \{[1, 2], [2, 3]\}$, $\delta([1, 2]) = 1$, $\delta([2, 3]) = 2$, $\lambda([1, 2]) = \lambda([2, 3]) = 1$ and $\pi(1) = \pi(2) = \pi(3) = 0$. It is easy to see that the unique $(1|1)$-centroid x is located on edge $[2, 3]$ with $d(1, x) = 1.25$.

Moreover, Spoerhase and Wirth (2009c) show that, on a chain network, the continuous version of the binary $(r|p)$-centroid problem with vertex demand only is NP-hard, while the discrete version is not. Hence, they conjecture that a polynomial finite dominating set is unlikely to exist for the continuous, binary $(r|p)$-centroid problem on general networks with vertex demand only. This implies that the existence of a finite dominating set is similarly unlikely in the case of vertex and edge demand. Hence, we leave the question of whether or not there exists a more general discretization result than the one given in Proposition 4.1.1 for future research and turn our attention to the discrete problem class.

4.2 An Efficient Algorithm on Tree Networks

In this section, we adapt the definition of ξ-bounding sets by Spoerhase and Wirth (2009c) to derive an efficient algorithm for the discrete, binary $(1|p)$-centroid problem on tree networks with vertex and edge demand. Let $N = (V, E, \lambda)$ be a tree network. Add an artificial vertex $n+1$ with $\pi(n+1) = 0$ as the root of N and connect it to an arbitrary vertex $s \in V$ by an artificial edge $[s, n+1]$ with $\lambda([s, n+1]) = \infty$ and $\delta([s, n+1]) = 0$.

Definition 4.2.1 (Spoerhase and Wirth (2009c)). Let $X \subseteq V \setminus \{n+1\}$ be a vertex subset and $0 \leq \hat{\xi} \leq \xi(N)$. Set X is called *$\hat{\xi}$-bounding* if

1. $W_F^*(Y_1(X)|X) \leq \hat{\xi}$ and
2. $\forall\, x \in X$ with father x' we have $W_F^*(Y_1((X \setminus \{x\}) \cup \{x'\})|(X \setminus \{x\}) \cup \{x'\}) > \hat{\xi}$.

Additionally, we define:

Definition 4.2.2. Let $N = (V, E, \lambda)$ be a tree network. Furthermore, let $X \subseteq V$, $v \in V$ and u be the father of v, $\{u, v\} \cap X = \emptyset$. Then the *$X$-subtracted subtree rooted in u* is the subnetwork on the vertex set

$$V_{X_s} := \{u\} \cup \left\{ V_{T_v} \setminus \bigcup_{i \in \{X \cap V_{T_v}\}} V'_{T_i} \right\}.$$

Before describing the algorithm itself in Section 4.2.2, we will introduce the basic ideas by restricting our attention to chain networks in Section 4.2.1. We impose an additional assumption on the networks under consideration, i.e. we assume $\lambda : E \to \mathbb{Q}^+$, $\delta : E \to \mathbb{Q}_0^+$ and $\pi : V \to \mathbb{Q}_0^+$. As a direct consequence, we may assume without loss of generality that the edge lengths, demand densities and vertex demands are integer numbers for the remainder of this chapter: Let all edge lengths, demand densities and vertex demands be expressed as fractions of two integers and define c to be the least common multiple of their denominators. Then we can transform N into a network $N' = (V, E, \lambda')$ with $\lambda'(uv) = c\lambda(uv)$, $\delta'(uv) = c\delta(uv)$ for all $[u, v] \in E$ and $\pi'(u) = c^2\pi(u)$ for all $u \in V$ without changing the ratio of the market shares of leader and follower for any feasible location setting.

4.2.1 Chain Networks

Consider a chain network $N = (V, E, \lambda)$ with vertex set $V = \{u_1, ..., u_n\}$ and edge set $E = \{[u_i, u_{i+1}] | i = 1, ..., n-1\}$ (see Figure 4.4). Global variables x and t allow the definition of any point of the chain network. For the sake of notational convenience, we will denote the vertices by natural numbers, i.e. i instead of u_i, whenever possible. The distinction of vertex numbers and distinct values of the global variables x and t will, in general, become clear from the context.

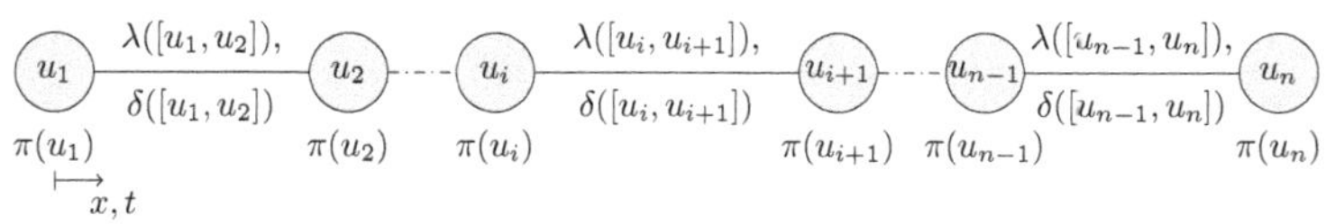

Fig. 4.4: Chain network

Consider the case $p = 1$. Obviously, if the leader locates at vertex u_i, an optimal solution to the medianoid problem will always exist at a neighbored vertex, i.e. u_{i-1} or u_{i+1}. Thus, a first result is as follows.

Lemma 4.2.1. *$W_L^*(1|1) \geq 0.5 \cdot \xi(N)$ for any instance of the discrete, binary (1|1)-centroid problem on a chain network $N = (V, E, \lambda)$ with vertex and edge demand.*

Proof. Define $\delta([u_n, u_{n+1}]) := 0$, $\delta([u_0, u_1]) := 0$, $\pi(u_0) := 0$ and $\pi(u_{n+1}) := 0$. Then there always exists

- at least one point (case 1) $x = \hat{x} \notin V$ of the chain network such that

$$\sum_{i=1}^{z-1}(\pi(u_i) + \lambda([u_i, u_{i+1}])\delta([u_i, u_{i+1}])) + \pi(u_z) \\ + (\hat{x} - d(1, u_z))\delta([u_z, u_{z+1}]) = 0.5 \cdot \xi(N),$$

  where $z := \max\{i | u_i \in V, d(1, u_i) \leq \hat{x}\}$,

- or (case 2) at least one vertex $u_{z'} \in V$ of the chain network such that

$$\sum_{i=1}^{z'-1}(\pi(u_i) + \lambda([u_i, u_{i+1}])\delta([u_i, u_{i+1}])) = 0.5 \cdot \xi(N),$$

- or (case 3) exactly one vertex $u_{z''} \in V$ of the chain network such that

$$\sum_{i=1}^{z''-1}(\pi(u_i) + \lambda([u_i, u_{i+1}])\delta([u_i, u_{i+1}])) + \pi(u_z) \geq 0.5 \cdot \xi(N) \text{ and}$$
$$\sum_{i=1}^{z''-1}(\pi(u_i) + \lambda([u_i, u_{i+1}])\delta([u_i, u_{i+1}])) < 0.5 \cdot \xi(N).$$

We consider the three cases:

- case 1:

  1. We have $W_F(u_{z-1}|u_z) \leq 0.5 \cdot \xi(N)$ and $W_F(u_{z+2}|u_{z+1}) \leq 0.5 \cdot \xi(N)$ by definition of z.
  2. If we assume that $X_1 = \{u_z\}$ and $W_F(u_{z+1}|X_1) > 0.5 \cdot \xi(N)$ ($W_F(u_{z+1}|X_1) = 0.5 \cdot \xi(N)$), then it is immediately implied that $W_F(u_z|X_1' = \{u_{z+1}\}) < 0.5 \cdot \xi(N)$ ($W_F(u_z|X_1' = \{u_{z+1}\}) = 0.5 \cdot \xi(N)$), so that the leader may choose X_1'. If we assume otherwise that $X_1 = \{u_z\}$ and $W_F(u_{z+1}|X_1) \leq 0.5 \cdot \xi(N)$, then the leader may choose X_1.

3. The same reasoning as in the previous subcase holds for assuming $X_1 = \{u_{z+1}\}$ and $W_F(u_z|X_1) > 0.5 \cdot \xi(N)$ $(W_F(u_z|X_1) = 0.5 \cdot \xi(N))$.

- case 2:
  We have $W_F(u_{z'-1}|u_{z'}) \leq 0.5 \cdot \xi(N)$ and $W_F(u_{z'+1}|u_{z'}) \leq 0.5 \cdot \xi(N)$ by definition of $u_{z'}$.

- case 3:
  We have $W_F(u_{z''-1}|u_{z''}) < 0.5 \cdot \xi(N)$ and $W_F(u_{z''+1}|u_{z''}) \leq 0.5 \cdot \xi(N)$ by definition of $u_{z''}$.

The case differentiation proves the claim. □

Hence, we can immediately conclude the following theorem.

Theorem 4.2.1. *The problem of finding a discrete, binary* (1|1)*-centroid of a chain network* $N = (V, E, \lambda)$ *with vertex and edge demand can be solved in linear time* $\mathcal{O}(n)$.

Proof. A straight forward algorithm, based on Lemma 4.2.1, is given below. It results in a (1|1)-centroid stored in the set X_1.

Algorithm 4.2.1 Discrete, binary (1|1)-centroid of chain network with vertex and edge demand

$X_1 := \emptyset$
$\xi_s := 0,\ i := 1$
while $\xi_s < \frac{1}{2}\xi(N)$ **do**
  $\xi_s := \xi_s + \pi(i)$
  if $\xi_s \geq \frac{1}{2}\xi(N)$ **then**
    $X_1 := \{i\}$
  else
    $\xi_s := \xi_s + \lambda([i, i+1])\delta([i, i+1])$
    if $\xi_s \geq \frac{1}{2}\xi(N)$ **then**
      if $\xi_s - \lambda([i, i+1])\delta([i, i+1]) \geq \frac{1}{2}(\xi(N) - \lambda([i, i+1])\delta([i, i+1]))$ **then**
        $X_1 := \{i\}$

        else
            $X_1 := \{i+1\}$
        end if
    end if
end if
$i := i+1$
end while

□

It will prove useful to design an efficient algorithm to determine a discrete, binary $(1|X_2)$-medianoid of a chain network $N = (V, E, \lambda)$ with vertex and edge demand (cf. also Dasci et al., 2002). Without loss of generality, let the leader's locations be the vertices j and k with $j \leq k$. Then a rather naive approach of time complexity $\mathcal{O}(n^2)$ is presented in Algorithm 4.2.2. It determines (not necessarily all) $(1|X_2)$-medianoids, stored in Y, as well as the corresponding market share of the follower, ξ_s^*.

Algorithm 4.2.2 Discrete, binary $(1|X_2)$-medianoid of chain network with vertex and edge demand

$Y := \emptyset$, $\lambda_s := 0$, $\xi_s := 0$, $\xi_s^* := 0$, $l := 1$
for all $i := j-1$ **to** $k+1$, $i \neq j$, $i \neq k$, $i \neq 0$, $i \neq n+1$ **do**
    $\xi_s := 0$
    if $i < j$ **then**
        $\lambda_s := 0$, $l := 1$
        while $\lambda_s < \frac{1}{2}(d(1,i) + d(1,j))$ **do**
            if $\lambda_s + \lambda([l, l+1]) < \frac{1}{2}(d(1,i) + d(1,j))$ **then**
                $\xi_s := \xi_s + \pi(l) + \lambda([l, l+1])\delta([l, l+1])$
            else
                $\xi_s := \xi_s + \pi(l) + \delta([l, l+1])(\frac{1}{2}(d(1,i) + d(1,j)) - \lambda_s))$
            end if
            $\lambda_s := \lambda_s + \lambda([l, l+1])$, $l := l+1$
        end while
    end if
    if $j < i < k$ **then**

$\lambda_s := d(1,j)$, $l := j$
while $\lambda_s < \frac{1}{2}(d(1,i) + d(1,k))$ **do**
if $\lambda_s+\lambda([l,l+1]) > \frac{1}{2}(d(1,i)+d(1,j))$ and $\lambda_s \leq \frac{1}{2}(d(1,i)+d(1,j))$ **then**
$\xi_s := \xi_s + \pi(l+1) + \delta([l,l+1])(\lambda_s + \lambda([l,l+1]) - \frac{1}{2}(d(1,i) + d(1,j)))$
end if
if $\lambda_s > \frac{1}{2}(d(1,i)+d(1,j))$ and $\lambda_s+\lambda([l,l+1]) < \frac{1}{2}(d(1,i)+d(1,k))$ **then**
$\xi_s := \xi_s + \pi(l+1) + \delta([l,l+1])\lambda([l,l+1])$
end if
if $\lambda_s + \lambda([l,l+1]) \geq \frac{1}{2}(d(1,i) + d(1,k))$ **then**
$\xi_s := \xi_s + \delta([l,l+1])(\frac{1}{2}(d(1,i) + d(1,k)) - \lambda_s)$
end if
$\lambda_s := \lambda_s + \lambda([l,l+1])$, $l := l+1$
end while
end if
if $i > j$ and $i > k$ **then**
$\lambda_s := d(1,k)$, $l := k$
while $\lambda_s < d(1,n)$ **do**
if $\lambda_s+\lambda([l,l+1]) > \frac{1}{2}(d(1,i)+d(1,k))$ and $\lambda_s \leq \frac{1}{2}(d(1,i)+d(1,k))$ **then**
$\xi_s := \xi_s + \pi(l+1) + \delta([l,l+1])(\lambda_s + \lambda([l,l+1]) - \frac{1}{2}(d(1,i) + d(1,k)))$
end if
if $\lambda_s > \frac{1}{2}(d(1,i) + d(1,k))$ **then**
$\xi_s := \xi_s + \pi(l+1) + \delta([l,l+1])\lambda([l,l+1])$
end if
$\lambda_s := \lambda_s + \lambda([l,l+1])$, $l := l+1$
end while
end if
if $\xi_s > \xi_s^*$ **then**
$Y := \{i\}$, $\xi_s^* := \xi_s$

  else if $\xi_s = \xi_s^*$ **then**
    $Y := Y \cup \{i\}$
  end if
end for
if $Y = \emptyset$ **then**
  $Y := Y \cup \{1\}$
end if

Observe that Algorithm 4.2.2 may consider some vertex and edge demands of the chain network multiple times. We can remove those redundant operations by defining

$$f(x) := \sum_{i \in V, d(1,i) < x} \pi(i) + \int_0^x \delta(t)dt, \tag{4.29}$$

where $\delta(t) := 0$ for all $t < 0$ and $t > d(1,n)$.

This results in

$$f(x) = \begin{cases} 0 & \text{if } x \le 0, \\ \pi(1) + \delta([1,2])x & \text{if } 0 < x \le d(1,2), \\ \vdots & \vdots \\ \sum_{i=1}^{u-2} [\delta([i,i+1])\lambda([i,i+1]) + \pi(i)] & \text{if } x > d(1,u-1) \\ \quad + \pi(u-1) & \text{and } x \le d(1,u), \\ \quad + \delta([u-1,u])(x - d(1,u-1)) & \\ \vdots & \vdots \\ \sum_{i=1}^{n-2} [\delta([i,i+1])\lambda([i,i+1]) + \pi(i)] & \text{if } x > d(1,n-1) \\ \quad + \pi(n-1) & \text{and } x \le d(1,n), \\ \quad + \delta([n-1,n])(x - d(1,n-1)) & \\ \sum_{i=1}^{n-1} [\delta([i,i+1])\lambda([i,i+1]) & \text{if } x > d(1,n). \\ \quad + \pi(i)] + \pi(n) & \end{cases} \tag{4.30}$$

Note that, after applying a simple preprocessing procedure of linear time complexity, one can evaluate $f(x)$ in constant time for any x. Figure 4.5 gives an example, where the labels of edges $[u, v]$ correspond to $(\lambda([u, v]), \delta([u, v]))$.

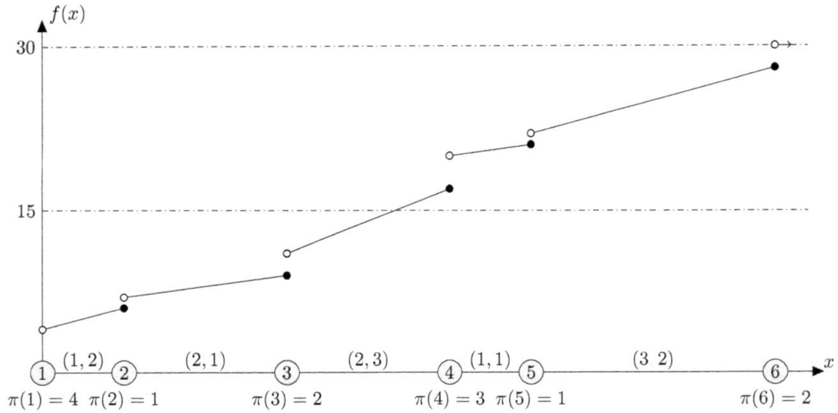

Fig. 4.5: Function $f(x)$

Let the leader's locations be the vertices j and k with $j \leq k$ and let the follower's location be vertex $i \neq j, k$. Then we can easily determine an open interval (a, b) with $a < b$ and $a, b \in [0, d(1, n)]$ of customers, who accommodate their demand at the follower's facility:

$$a = \begin{cases} 0 & \text{if } i < j, \\ \frac{1}{2}(d(1,j) + d(1,i)) & \text{if } j < i < k, \\ \frac{1}{2}(d(1,k) + d(1,i)) & \text{if } i > k, \end{cases} \tag{4.31}$$

$$b = \begin{cases} \frac{1}{2}(d(1,j) + d(1,i)) & \text{if } i < j, \\ \frac{1}{2}(d(1,k) + d(1,i)) & \text{if } j < i < k, \\ d(1,n) & \text{if } i > k. \end{cases} \tag{4.32}$$

The customers located at a and b have to be considered separately, so that the follower's market share can be calculated as follows:

$$W_F(i|j,k) = \begin{cases} f(b) - f(a) & \text{if } i < j, \\ f(b) - f(a) - \pi(a) & \text{if } j < i < k, \\ f(b) - f(a) - \pi(a) + \pi(n) & \text{if } i > k. \end{cases} \tag{4.33}$$

Thus, we can state:

Lemma 4.2.2. *The problem of finding a discrete, binary* $(1|X_2)$*-medianoid of a chain network* $N = (V, E, \lambda)$ *with vertex and edge demand can be solved in linear time* $\mathcal{O}(n)$.

Proof. An algorithm is as follows. It determines (not necessarily all) $(1|X_2)$-medianoids, stored in Y, as well as the corresponding market share of the follower, ξ_s^*.

Algorithm 4.2.3 Discrete, binary $(1|X_2)$-medianoid of chain network with vertex and edge demand

$Y := \emptyset$, $\xi_s := 0$, $\xi_s^* := 0$
for all $i := j - 1$ **to** $k + 1$, $i \neq j$, $i \neq k$, $i \neq 0$, $i \neq n + 1$ **do**
  $\xi_s := W_F(i|j,k)$ (Equations (4.31)-(4.33))
  if $\xi_s > \xi_s^*$ **then**
    $Y := \{i\}$, $\xi_s^* := \xi_s$
  else if $\xi_s = \xi_s^*$ **then**
    $Y := Y \cup \{i\}$
  end if
end for
if $Y = \emptyset$ **then**
  $Y := \{1\}$
end if

□

We are now able to derive an efficient algorithm for the centroid problem by combining Algorithm 4.2.3 and the definition of $\hat{\xi}$-bounding sets, presented by Spoerhase and Wirth (2009c) for the case of vertex demand only. To this end we add an artificial vertex $n+1$ with $\pi(n+1) = 0$ and an artificial edge $[n, n+1]$ with $\lambda([n, n+1]) = \infty$ and $\delta([n, n+1]) = 0$ to the chain network.

Lemma 4.2.3. *Let $0 < p \leq n$ with $p \in \mathbb{N}^+$ and $0 \leq \hat{\xi} \leq \xi(N)$ with $\hat{\xi} \in \mathbb{R}_0^+$. If, for a chain network $N = (V, E, \lambda)$ with vertex and edge demand, there exists a $\hat{\xi}$-bounding set with at most p elements, we can determine one such set (or decide that there is no such set) in $\mathcal{O}(n^2)$ time.*

Proof. Observe that the cases $\hat{\xi} = 0$ and $\hat{\xi} = \xi(N)$ are trivial, i.e. we can easily decide if a $\hat{\xi}$-bounding set exists and if so, it is a simple matter to find such a set. Hence, we assume $0 < \hat{\xi} < \xi(N)$ in the following.

Algorithm 4.2.4 determines a set X of at most p elements that form a $\hat{\xi}$-bounding set in the proposed running time. If there is no such set, the algorithm terminates with $X = \emptyset$.

Algorithm 4.2.4 $\hat{\xi}$-bounding set X with $|X| \leq p$ of chain network with vertex and edge demand

$X := \emptyset$, $\xi_s := \pi(1) + \frac{1}{2}\delta([1,2])\lambda([1,2])$, $i := 2$
while $\xi_s \leq \hat{\xi}$ **and** $i \leq n-1$ **do**
    $i := i + 1$
    $\xi_s := \xi_s + \frac{1}{2}\delta([i-2, i-1])\lambda([i-2, i-1]) + \pi(i-1) + \frac{1}{2}\delta([i-1, i])\lambda([i-1, i])$
end while
if $i = n$ **and** $\xi_s \leq \hat{\xi}$ **then**
    $X := X \cup \{i\}$
else
    $X := X \cup \{i-1\}$
end if
while $|X| \leq p$ **and** $i \leq n$ **do**
    $\xi_s := 0$
    while $\xi_s \leq \hat{\xi}$ **and** $i \leq n$ **do**
        $i := i + 1$
        Call Algorithm 4.2.3 on the subnetwork N' with vertex set $V' = \{\max\{l | l \in X\}, ..., i\}$ and $j = \max\{l | l \in X\}$, $k = i$. It results in ξ_s^* and a set Y.

```
        ξ_s := ξ_s^*
    end while
    if ξ_s > ξ̂ then
        X := X ∪ {i − 1}
    end if
end while
if |X| = p + 1 then
    X := ∅
end if
```

□

Obviously, the set of all $(1|p)$-centroids of a chain network with follower gain $\hat{\xi}$ contains all $\hat{\xi}$-bounding sets with at most p elements. Thus, we may state the main result of this section:

Theorem 4.2.2. *The problem of finding a discrete, binary $(1|p)$-centroid of a chain network $N = (V, E, \lambda)$ with vertex and edge demand can be solved with time complexity $\mathcal{O}(n^2 \log \xi(N))$.*

Proof. We apply bisection on the interval $[0, \xi(N)]$ to find the smallest value $\hat{\xi}$ such that a $\hat{\xi}$-bounding set with at most p elements exists (cf. Spoerhase and Wirth, 2009c). Observe that (due to the additional assumption in Section 4.2) we can terminate the bisection method when $|\xi_{best} - \xi_{nobs}| < 0.5$, where ξ_{best} corresponds to the follower's market share in the best feasible solution found so far, and ξ_{nobs} corresponds to the largest demand level in an iteration of the bisection method that resulted in the nonexistence of a ξ_{nobs}-bounding set with at most p elements. This results in the proposed running time. □

4.2.2 General Tree Networks

Suppose that the given network $N = (V, E, \lambda)$ is a tree network with vertex set $V = \{u_1, ..., u_n\}$ and edge set E. We will denote the vertices by natural numbers, i.e. i instead of u_i, whenever possible. As in the case of a chain network, the set of all $(1|p)$-centroids of a tree network with follower gain $\hat{\xi}$ contains all $\hat{\xi}$-bounding sets with at most p elements. As we will show in the following, we can extend the ideas presented in Section 4.2.1 to determine a discrete, binary $(1|p)$-centroid of a tree network with vertex and edge demand.

Add an artificial vertex $n+1$ with $\pi(n+1) = 0$ as the root of N and connect it to an arbitrary vertex $s \in V$ by an artificial edge $[s, n+1]$ with $\lambda([s, n+1]) = \infty$ and $\delta([s, n+1]) = 0$. A $\hat{\xi}$-bounding set with at most p elements can be determined by performing a depth first search traversal of the tree network (cf. Spoerhase and Wirth, 2009c).

Algorithm 4.2.5 $\hat{\xi}$-bounding set X with $|X| \leq p$ of tree network with vertex and edge demand

---

1: $X := \emptyset$
2: Perform a depth first search traversal of the tree network, starting at the artificial vertex $n+1$. Whenever the search tracks back from vertex v to vertex u, solve a $(1|X_{\hat{p}})$-medianoid problem on the X-subtracted subtree rooted in u with $X_{\hat{p}} = (X \cap V_{X_s}) \cup \{u\}$. If the follower's optimal market share exceeds $\hat{\xi}$, set $X = X \cup \{v\}$.
3: **if** $|X| > p$ **then**
4: $\quad X := \emptyset$
5: **end if**

Suppose for now that we have an appropriate algorithm to solve discrete, binary $(1|X_p)$-medianoid problems on tree networks at hand. Then the main idea of how to find a discrete, binary $(1|p)$-centroid is in analogy to Section 4.2.1, i.e. we apply bisection on the interval $[0, \xi(N)]$ to find the smallest value $\hat{\xi}$ such that a $\hat{\xi}$-bounding set with at most p elements exists (cf. Spoerhase and Wirth, 2009c). The following example illustrates this basic idea. Suppose we want to solve a $(1|2)$-centroid problem on the tree network depicted in Figure 4.6, with all edge lengths, demand densities and vertex demands equal to one. We arbitrarily choose $s = 1$. Figure 4.7 highlights and numbers all the subtrees that we will have to consider in the solution process. Shaded vertices represent leader's facilities. Table 4.3 lists corresponding optimal solutions to the medianoid problems. Table 4.4 walks through the solution process as described above. The termination criterion, $\hat{\xi}_7 - \hat{\xi}_6 < 0.5$, is applicable due to the integrality of the edge lengths, vertex demands and demand densities (see below).

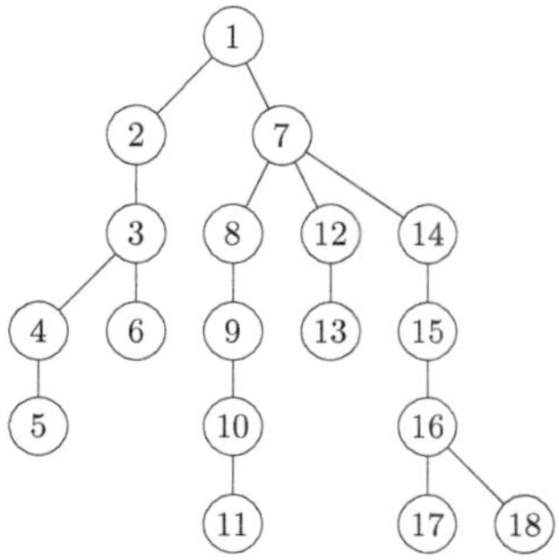

Fig. 4.6: Example - tree network

Tab. 4.3: Example - optimal solutions to the medianoid problems

Subtree	1	2	3	4	5	6	7	8	9	10
Opt. solution	{5}	{4}	{6}	{3}	{2}	{11}	{10}	{9}	{8}	{13}
Follower gain	1.5	3.5	1.5	7.5	9.5	1.5	3.5	5.5	7.5	1.5

Subtree	11	12	13	14	15	16	17	18	19	20
Opt. solution	{12}	{17}	{18}	{16}	{15}	{14}	{7}	{s}	{7}	{s}
Follower gain	3.5	1.5	1.5	5.5	7.5	9.5	23.5	11.5	14	1

As mentioned in Chapter 2.3, Dasci et al. (2002) propose an $\mathcal{O}(nm^2)$ algorithm for the continuous version of the $(1|X_p)$-medianoid problem on general networks with vertex and edge demand under a different tie breaking criterion than the one considered in this chapter. However, they do not explicitly present specific results on tree networks or on the discrete problem class. The $\mathcal{O}(rn^2)$ algorithm of Megiddo et al. (1983) for the continuous $(r|X_p)$-medianoid problem on tree networks with vertex demand only applies to the discrete version of the problem as well. It has the same time complexity in this latter case. Spoerhase and Wirth (2009a) improve this result by presenting an $\mathcal{O}(n(\log n)^2/\log\log n)$ algorithm for the case $r = 1$. However, as can be seen from Figure 4.7, the medianoid problems to be solved in our case are of a special structure: the leader's facilities are solely located in leaves and roots of subtrees of the original network. We will now exploit

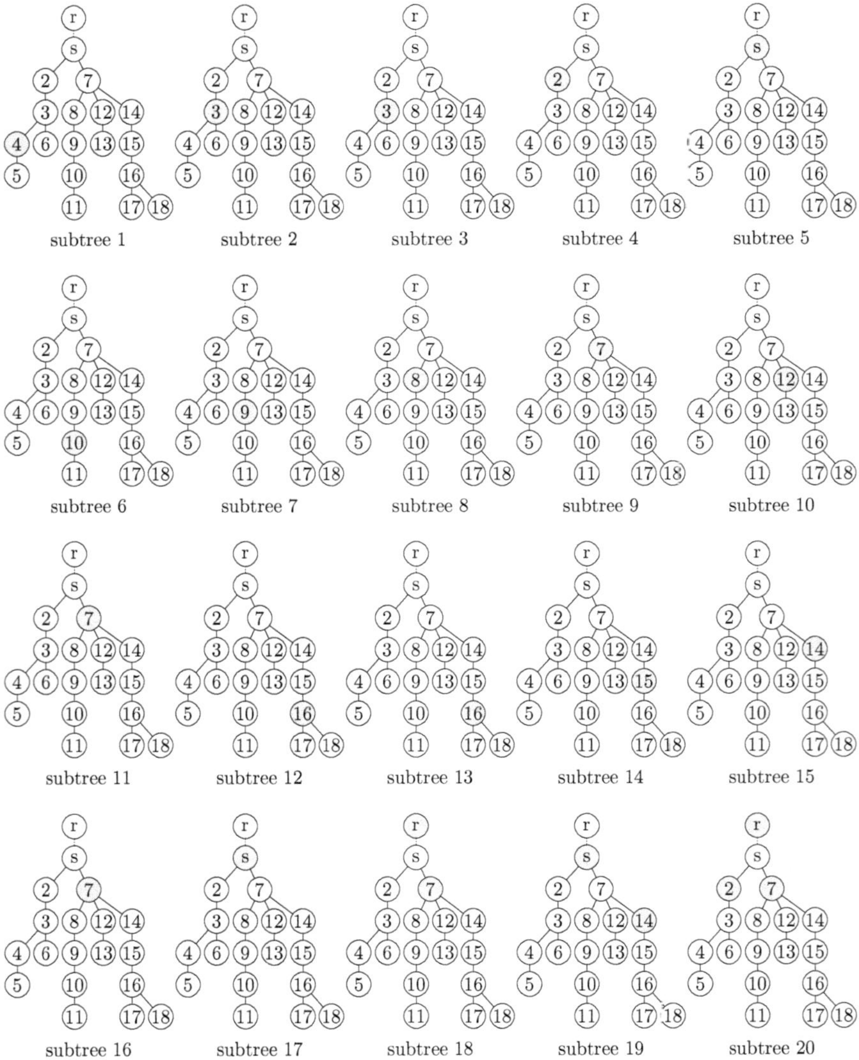

Fig. 4.7: Example - subtrees

Tab. 4.4: Example - solving the centroid problem

Bisection method	Algorithm 4.2.5: subnetworks under consideration	Bounding set
$\hat{\xi}_1 = 0.5\xi(N) = 17.5$	1, 2, 3, 4, 5, 6, 7, 8, 9, 10, 11, 12, 13, 14, 15, 16, 17 $\to X = \{7\}$ 18	$X = \{7\}$
$\hat{\xi}_2 = 8.75$	1, 2, 3, 4, 5 $\to X = \{2\}$ 6, 7, 8, 9, 10, 11, 12, 13, 14, 15, 16 $\to X = \{2, 14\}$ 19 $\to X = \{2, 7, 14\}$ 20 $\to \|X\| > 2$	$X = \emptyset$
$\hat{\xi}_3 = 13.125$	1, 2, 3, 4, 5, 6, 7, 8, 9, 10, 11, 12, 13, 14, 15, 16, 17 $\to X = \{7\}$ 18	$X = \{7\}$
$\hat{\xi}_4 = 10.9375$	1, 2, 3, 4, 5, 6, 7, 8, 9, 10, 11, 12, 13, 14, 15, 16, 17 $\to X = \{7\}$ 18 $\to X = \{1, 7\}$	$X = \{1, 7\}$
$\hat{\xi}_5 = 9.84375$	1, 2, 3, 4, 5, 6, 7, 8, 9, 10, 11, 12, 13, 14, 15, 16, 17 $\to X = \{7\}$ 18 $\to X = \{1, 7\}$	$X = \{1, 7\}$
$\hat{\xi}_6 = 9.296875$	1, 2, 3, 4, 5 $\to X = \{2\}$ 6, 7, 8, 9, 10, 11, 12, 13, 14, 15, 16 $\to X = \{2, 14\}$ 19 $\to X = \{2, 7, 14\}$ 20 $\to \|X\| > 2$	$X = \emptyset$
$\hat{\xi}_7 = 9.5703125$	1, 2, 3, 4, 5, 6, 7, 8, 9, 10, 11, 12, 13, 14, 15, 16, 17 $\to X = \{7\}$ 18 $\to X = \{1, 7\}$	$X = \{1, 7\}$
	STOP ($\hat{\xi}_7 - \hat{\xi}_6 < 0.5$) $X_2 = \{1, 7\}$ is a (1|2)-centroid	

this observation and describe an algorithm to optimally solve those special structure $(1|X_{\hat{p}})$-medianoid problems in $\mathcal{O}(\hat{n}\hat{p})$ time:

Lemma 4.2.4. *Let $\hat{N} = (\hat{V}, \hat{E}, \hat{\lambda})$, $|\hat{V}| = \hat{n}$, be a rooted tree network with vertex and edge demand.*[3] *Moreover, let $X_{\hat{p}} \subseteq \hat{L} \cup \{\hat{r}\}$, $X_{\hat{p}} \cap \{\hat{r}\} \neq \emptyset$, where $\hat{L}$ denotes the set of leaves and $\hat{r}$ corresponds to the root of the tree network with degree 1. Then the problem of finding a discrete, binary $(1|X_{\hat{p}})$-medianoid can be solved with time complexity $\mathcal{O}(\hat{n}\hat{p})$.*

The algorithm is composed of six stages.

1. Collapse, $\mathcal{O}(\hat{n})$

3 For notational convenience, we will denote the vertex and edge demand mappings in $\hat{N}$ (and $\hat{N}'$, see below) by π and δ as in the original network.

2. Label, $\mathcal{O}(\hat{n})$

3. Rearrange, $\mathcal{O}(\hat{p}^2)$

4. Construct chain functions, $\mathcal{O}(\hat{n}\hat{p})$

5. Calculate distances of facilities in X_p, $\mathcal{O}(\hat{p}^2)$

6. Evaluate, $\mathcal{O}(\hat{n}\hat{p})$

In the following we will describe each of the six stages in detail. We denote the elements of the set $X_{\hat{p}}$ by x_i, $i = 1, ..., \hat{p}$, where, without loss of generality, x_1 corresponds to the leader's facility located in the root $\hat{r}$.

Collapse

Consider a leaf $x \notin X_{\hat{p}}$ and let $x' \neq \hat{r}$ be the father of x. It is easy to see that $W_F(x|X_{\hat{p}}) \leq W_F(x'|X_{\hat{p}})$. Therefore, we may construct an auxiliary tree network $\hat{N}' = (\hat{V}', \hat{E}', \hat{\lambda}')$ by merging x and x' to define a vertex $v_{x,x'}$ with $\pi(v_{x,x'}) = \pi(x) + \pi(x') + \hat{\lambda}([x', x])\delta([x', x])$ to replace x, x' and $[x', x]$. Any $(1|X_{\hat{p}})$-medianoid on $\hat{N}'$ corresponds to a $(1|X_{\hat{p}})$-medianoid on $\hat{N}$. The transformation is straight forward: Let $v_{x,x'}$ be an optimal solution to the $(1|X_{\hat{p}})$-medianoid problem on $\hat{N}'$; then x' is the corresponding $(1|X_{\hat{p}})$-medianoid on $\hat{N}$. This idea of merging vertices can be repeated until every leaf of the resulting auxiliary tree network is either a child of $\hat{r}$ or an element of $X_{\hat{p}}$. This can be achieved by performing a depth first search traversal of the tree network of time complexity $\mathcal{O}(\hat{n})$. An example with $X_{\hat{p}} = \{x_1, x_2, x_3, x_4, x_5\}$ is given in Figure 4.8. Vertex labels outside of vertices v correspond to $\pi(v)$; labels of edges $[u, v]$ correspond to $(\hat{\lambda}([u, v]), \delta([u, v]))$. Since vertex 13 is a leaf without being a leader's location, we may merge vertices 13 and 12. The resulting vertex can subsequently be merged with vertices 11 and 6.

Observe that for $\hat{p} \leq 2$ we are left with a chain network (see Section 4.2.1). Therefore, in what follows, we will restrict our attention to the nontrivial case $\hat{p} > 2$. Moreover, to ease notation, we will omit the superscript $'$ when referring to the collapsed tree network and its vertex set, i.e. we will keep using the notation that has been introduced in Lemma 4.2.4.

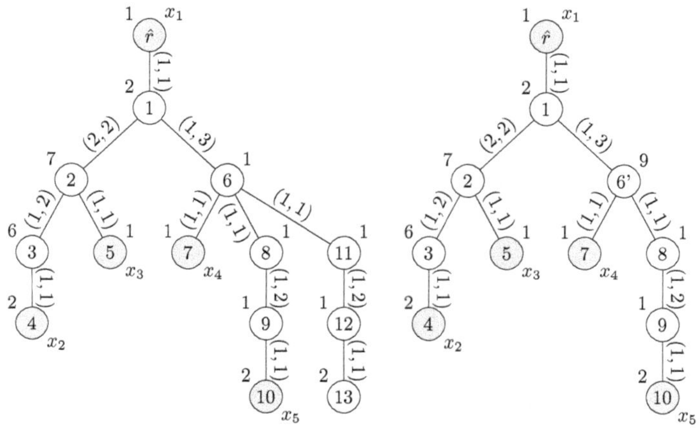

(a) Original tree network $\hat{N}$ *(b) Auxiliary tree network* $\hat{N}'$

Fig. 4.8: Collapsing stage

Label

It is well known that the vertices of a tree network can be labeled in linear time such that hereafter the nearest common ancestor of any pair of vertices as well as their distance can be computed in constant time. See, for example, Alstrup et al. (2004) and the references therein.

Rearrange

This stage aims at finding a permutation $[x_{\mu_1}, ..., x_{\mu_{\hat{p}-1}}]$ of the elements x_i, $i = 2, ...\hat{p}$. Roughly speaking, this permutation is such that the customers located on the unique path that connects a vertex that corresponds to a superordinated element of the permutation to the root are guaranteed not to accommodate their demand at a subordinated leader's facility of the permutation (or to be indifferent about accommodating their demand at one of the two corresponding leader's facilities). Additionally, in combination with the subsequent stages of the algorithm, the permutation eventually guarantees that each customer accommodates his demand at a single facility by grouping the leader's facilities according to their nearest common ancestors. The permutation is being determined by applying a three stage algorithm:

1. Calculate $d(\hat{r}, x_i)$ for all $i = 2, ...\hat{p}$ by use of the vertex labels (Stage 2) (time complexity $\mathcal{O}(\hat{p})$).

2. Sort the elements x_i, $i = 2, ...\hat{p}$, in the order of increasing distances $d(\hat{r}, x_i)$. We may, for example, apply mergesort (time complexity $\mathcal{O}(\hat{p} \log \hat{p})$, cf. Cormen et al., 2001). The sorting algorithm results in an array L of length $\hat{p} - 1$ with the smallest element in the first position, denoted by $L[1]$.

3. Call Algorithm 4.2.6 to determine the desired permutation P. We denote the i-th position of the permutation P by $P[i]$. Using a linked list data structure, the algorithm has time complexity $\mathcal{O}(\hat{p}^2)$.

Algorithm 4.2.6 Generate rearranged facilities

```
initialize P
pos := 0, lastpos := 2, depth := 0
P[1] := L[1], P[2] := L[2]
for all i := 3 to p̂ − 1 do
   pos := 0
   depth := d(r̂, nca(L[i], P[1]))
   for all j := 2 to lastpos do
      if d(r̂, nca(L[i], P[j])) > depth then
         depth := d(r̂, nca(L[i], P[j]))
         pos := j + 1
      end if
   end for
   if pos = 0 then
      if d(r̂, nca(L[i], P[1])) > d(r̂, nca(L[i], P[2])) then
         pos := 2
      else
         pos := lastpos + 1
      end if
   end if
```

```
20:     P[pos] := L[i]
21:     lastpos := lastpos + 1
22: end for
```

Let us denote the unique path that connects the vertex that corresponds to x_{μ_j} to the root $\hat{r}$ by c_{μ_j} for each $j = 1, ..., \hat{p} - 1$. Then the three stage rearranging process is such that for any pair $k, l \in \{1, ..., \hat{p} - 1\}$ with $k < l$ there exists a μ_m, $m \leq k$ such that $d(x_{\mu_m}, z) \leq d(x_{\mu_l}, z)$ for all points z on c_{μ_k}, even if $d(x_{\mu_l}, \hat{r}) < d(x_{\mu_k}, \hat{r})$.

We conclude by revisiting the example of Figure 4.8b in Table 4.5.

Tab. 4.5: Rearranging stage for the example in Fig. 4.8b

Steps 1 and 2:	distances:	sorted distances:
	i: 2 3 4 5	i: 4 3 2 5
	$d(\hat{r}, x_i)$: 5 4 3 5	$d(\hat{r}, x_i)$: 3 4 5 5
Alg. 4.2.6, initialization:	$L = [x_4, x_3, x_2, x_5]$	$P = [x_4, x_3]$
Alg. 4.2.6, $i = 3$:	insert x_2:	
	$nca(x_4, x_2) = 1$, $d(\hat{r}, 1) = 1$	
	$nca(x_3, x_2) = 2$, $d(\hat{r}, 2) = 3$	$P = [x_4, x_3, x_2]$
Alg. 4.2.6, $i = 4$:	insert x_5:	
	$nca(x_4, x_5) = 6'$, $d(\hat{r}, 6') = 2$	
	$nca(x_3, x_5) = 1$, $d(\hat{r}, 1) = 1$	
	$nca(x_2, x_5) = 1$, $d(\hat{r}, 1) = 1$	$P = [x_4, x_5, x_3, x_2]$

When inserting x_2, the algorithm compares the "depth" of the nearest common ancestors of x_2 and the facilities x_4 and x_3 in the tree network. Since the latter nearest common ancestor is located deeper in the network, x_2 is inserted after x_3. Similarly, x_5 needs to be paired with facility x_4. Thus, the procedure terminates with $x_{\mu_2} = x_5$ and $x_{\mu_3} = x_3$ although $d(\hat{r}, x_3) = 4 < d(\hat{r}, x_5) = 5$ because x_4 will always be closer to any point on chain $[\hat{r}, 1, 6', 8, 9, 10]$ than x_3.

Construct chain functions

As in the case of chain networks (Section 4.2.1), we can now define "chain functions" $f_{\mu_i}(y_{\mu_i})$ for all $i = 1, ..., \hat{p} - 1$, where y_{μ_i} are local variables that correspond to the distance of a point on chain c_{μ_i} to the root $\hat{r}$ of the collapsed tree network. In the following we will denote the vertices of a chain c_{μ_i} in the sequence of increasing distance to the root $\hat{r}$ by $v^1_{\mu_i}, \ldots, v^{l_{\mu_i}+1}_{\mu_i}$, where l_{μ_i} is the number of edges on the unique path from $\hat{r}$ to x_{μ_i}. In contrast to Section 4.2.1 we will now have to assign each vertex/edge customer to a unique chain. We do so by considering the permutation P that results from the rearranging stage. For any $i > 1$, we denote the vertex $nca(x_{\mu_{i-1}}, x_{\mu_i})$ by $v^s_{\mu_i}$, where $1 \leq s \leq l_{\mu_i}$, and we define

$$f_{\mu_i}(y_{\mu_i}) := \begin{cases} 0 & \text{if } y_{\mu_i} \leq d(\hat{r}, v^s_{\mu_i}), \\ \delta([v^s_{\mu_i}, v^{s+1}_{\mu_i}])(y_{\mu_i} - d(\hat{r}, v^s_{\mu_i})) & \text{if } y_{\mu_i} > d(\hat{r}, v^s_{\mu_i}) \text{ and } y_{\mu_i} \leq d(\hat{r}, v^{s+1}_{\mu_i}), \\ \vdots & \vdots \\ \sum_{j=s}^{u-1} [\delta([v^j_{\mu_i}, v^{j+1}_{\mu_i}])\hat{\lambda}([v^j_{\mu_i}, v^{j+1}_{\mu_i}]) + \pi(v^j_{\mu_i})] - \pi(v^s_{\mu_i}) + \pi(v^u_{\mu_i}) + \delta([v^u_{\mu_i}, v^{u+1}_{\mu_i}])(y_{\mu_i} - d(\hat{r}, v^u_{\mu_i})) & \text{if } y_{\mu_i} > d(\hat{r}, v^u_{\mu_i}) \text{ and } y_{\mu_i} \leq d(\hat{r}, v^{u+1}_{\mu_i}), \\ \vdots & \vdots \\ \sum_{j=s}^{l_{\mu_i}-1} [\delta([v^j_{\mu_i}, v^{j+1}_{\mu_i}])\hat{\lambda}([v^j_{\mu_i}, v^{j+1}_{\mu_i}]) + \pi(v^j_{\mu_i})] - \pi(v^s_{\mu_i}) + \pi(v^{l_{\mu_i}}_{\mu_i}) + \delta([v^{l_{\mu_i}}_{\mu_i}, v^{l_{\mu_i}+1}_{\mu_i}])(y_{\mu_i} - d(\hat{r}, v^{l_{\mu_i}}_{\mu_i})) & \text{if } y_{\mu_i} > d(\hat{r}, v^{l_{\mu_i}}_{\mu_i}) \text{ and } y_{\mu_i} \leq d(\hat{r}, v^{l_{\mu_i}+1}_{\mu_i}), \\ \sum_{j=s}^{l_{\mu_i}} [\delta([v^j_{\mu_i}, v^{j+1}_{\mu_i}])\hat{\lambda}([v^j_{\mu_i}, v^{j+1}_{\mu_i}]) + \pi(v^j_{\mu_i})] - \pi(v^s_{\mu_i}) + \pi(v^{l_{\mu_i}+1}_{\mu_i}) & \text{if } d(\hat{r}, v^{l_{\mu_i}+1}_{\mu_i}) < y_{\mu_i}. \end{cases} \tag{4.34}$$

For $i = 1$ we define $v^s_{\mu_1} := \hat{r}$ and $f_{\mu_1}(y_{\mu_1})$ in analogy to equation (4.30).

A simple $\mathcal{O}(\hat{n}\hat{p})$ procedure to construct the chain functions determines the chains c_{μ_i} for all $i = 1, \ldots, \hat{p} - 1$ by performing a depth first search traversal of the collapsed tree network and, afterwards, generates the needed coefficients by considering all edges of the collapsed tree network at most $\hat{p}$ times. After having constructed a chain function c_{μ_i}, we can evaluate it in constant time for a given y_{μ_i}.

For our example, the chain functions $f_{\mu_1}(y_{\mu_1}), ..., f_{\mu_4}(y_{\mu_4})$ for the permutation $P = [x_4, x_5, x_3, x_2]$ are depicted in Figure 4.9. Since, for instance, $nca(x_4, x_5) = 6'$, we have $f_{x_5}(y_{x_5}) = 0$ for all $y_{x_5} \leq d(\hat{r}, 6')$. Hence, the vertex demand of the vertices $\hat{r}$, 1 and $6'$ as well as the edge demand of the edges $[\hat{r}, 1]$ and $[1, 6']$ is uniquely assigned to chain c_{x_4}.

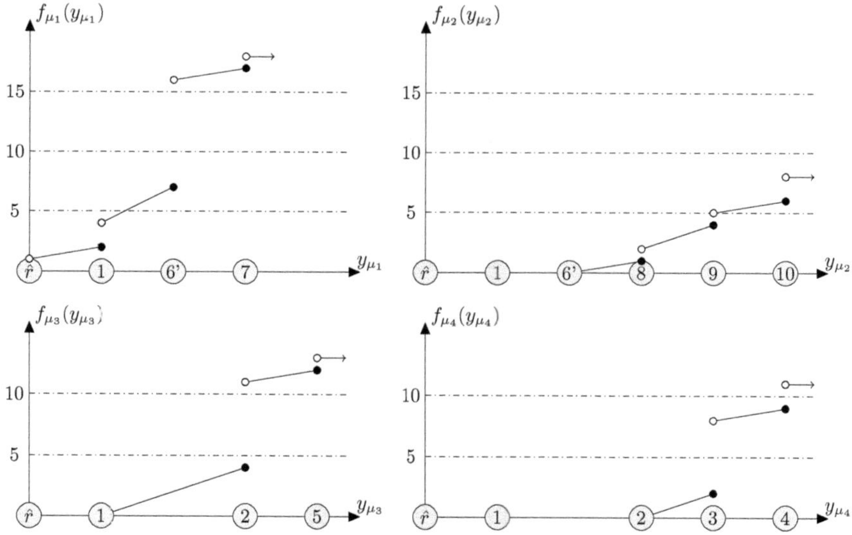

Fig. 4.9: Functions $f_{\mu_i}(y_{\mu_i})$, $i = 1, ..., 4$, for the example in Fig. 4.8b

Calculate distances of facilities in X_p

Determine

$$D(x_{\mu_i}) := \min\{d(\hat{r}, x_{\mu_i}), \min\{d(x_{\mu_i}, x_{\mu_j}) | j < i\}\} \tag{4.35}$$

for all $i = 2, ..., \hat{p} - 1$ and denote one of the corresponding vertices by $V(D(x_{\mu_i}))$. This can obviously be achieved in $\mathcal{O}(\hat{p}^2)$ time. Moreover, we equivalently define $D(x_{\mu_1}) := d(\hat{r}, x_{\mu_1})$ and $V(D(x_{\mu_1})) := \hat{r}$. The example is revisited in Table 4.6.

Tab. 4.6: $D(x_{\mu_i})$ and $V(D(x_{\mu_i}))$, $i = 1, ..., 4$, for the example in Fig. 4.8b

i	1	2	3	4
x_{μ_i}	x_4	x_5	x_3	x_2
$D(x_{\mu_i})$	3	4	4	3
$V(D(x_{\mu_i}))$	$\hat{r}$	7	$\hat{r}$	5

Evaluate

Consider a vertex $v \in \hat{V} \setminus X_{\hat{p}}$ of the collapsed tree network as a potential follower's location and let $k = \min\{i | v \text{ is a vertex on chain } c_{\mu_i}, i \in \{1, ..., \hat{p}-1\}\}$. Then we can easily determine an open interval $(a_{c_{\mu_j}}, b_{c_{\mu_j}})$ with $a_{c_{\mu_j}} \leq b_{c_{\mu_j}}$ of customers who accommodate their demand at v for each chain c_{μ_j}, $j = 1, ..., \hat{p} - 1$ in analogy to Section 4.2.1.

- For all c_{μ_j} with $j > k$:

$$a_{c_{\mu_j}} = 0, \tag{4.36}$$

$$b_{c_{\mu_j}} = \begin{cases} d(\hat{r}, x_{\mu_j}) - \frac{d(v, x_{\mu_j})}{2} & \text{if } d(v, x_{\mu_j}) < D(x_{\mu_j}), \\ 0 & \text{else.} \end{cases} \tag{4.37}$$

- For all c_{μ_j} with $j < k$, $j \neq 1$:

$$a_{c_{\mu_j}} = \begin{cases} d(\hat{r}, v) - \frac{d(V(D(x_{\mu_j})), v)}{2} & \text{if } d(\hat{r}, v) < d(\hat{r}, x_{\mu_j}) \\ & \quad \text{and } d(x_{\mu_j}, v) < D(x_{\mu_j}), \\ 0 & \text{else,} \end{cases} \tag{4.38}$$

$$b_{c_{\mu_j}} = \begin{cases} d(\hat{r}, x_{\mu_j}) - \frac{d(v, x_{\mu_j})}{2} & \text{if } d(\hat{r}, v) < d(\hat{r}, x_{\mu_j}) \\ & \quad \text{and } d(x_{\mu_j}, v) < D(x_{\mu_j}), \\ 0 & \text{else.} \end{cases} \tag{4.39}$$

- For c_{μ_k}:

$$a_{c_{\mu_k}} = d(\hat{r}, v) - \frac{d(V(D(x_{\mu_k})), v)}{2}, \tag{4.40}$$

$$b_{c_{\mu_k}} = \frac{d(\hat{r}, x_{\mu_k}) + d(\hat{r}, v)}{2}. \tag{4.41}$$

- For c_{μ_1}, if $k \neq 1$:

$$a_{c_{\mu_1}} = \begin{cases} \frac{d(\hat{r},v)}{2} & \text{if } d(\hat{r}, v) < d(\hat{r}, x_{\mu_1}) > d(v, x_{\mu_1}), \\ 0 & \text{else,} \end{cases} \tag{4.42}$$

$$b_{c_{\mu_1}} = \begin{cases} d(\hat{r}, x_{\mu_1}) - \frac{d(v,x_{\mu_1})}{2} & \text{if } d(\hat{r}, v) < d(\hat{r}, x_{\mu_1}) > d(v, x_{\mu_1}), \\ 0 & \text{else.} \end{cases} \tag{4.43}$$

Observe that there might exist points of the collapsed tree network that are contained in multiple intervals $(a_{c_{\mu_j}}, b_{c_{\mu_j}})$, $j \in \{1, ..., \hat{p}-1\}$. However, by definition of equation (4.34) and $f_{\mu_1}(y_{\mu_1})$, only one of the corresponding chain functions $f_{\mu_j}(y_{\mu_j})$ may have a value larger than zero at each of these points, so that the related demand will only be considered once in the subsequent calculation of the actual follower's market share $W_F(v|X_{\hat{p}})$. In analogy to Section 4.2.1 we get

$$W_F(v|X_{\hat{p}}) = \sum_{j=1}^{\hat{p}-1} W_F^{c_{\mu_j}}(v|X_{\hat{p}}), \tag{4.44}$$

where

$$W_F^{c_{\mu_j}}(v|X_{\hat{p}}) = \begin{cases} f_{\mu_j}(b_{c_{\mu_j}}) - f_{\mu_j}(a_{c_{\mu_j}}) & \text{if } d(\hat{r}, v^s_{\mu_j}) \geq a_{c_{\mu_j}}, \\ f_{\mu_j}(b_{c_{\mu_j}}) - f_{\mu_j}(a_{c_{\mu_j}}) - \pi(a_{c_{\mu_j}}) & \text{else,} \end{cases} \tag{4.45}$$

corresponds to the portion of the total follower's market share that is induced by chain c_{μ_j} for any $j \in \{1, ..., \hat{p}-1\}$.

Equations (4.36)-(4.45) can be evaluated in constant time for a given v, so that we are now able to determine a (special structure) discrete, binary

$(1|X_{\hat{p}})$-medianoid with time complexity $\mathcal{O}(\hat{n}\hat{p})$ as claimed in Lemma 4.2.4. The following algorithm stores $(1|X_{\hat{p}})$-medianoids in the set Y; the corresponding follower's market share equals W^*.

Algorithm 4.2.7 Evaluate

$Y := \emptyset$, $W := 0$, $W^* := 0$
for all $v \in \hat{V} \setminus X_{\hat{p}}$ **do**
  $W := 0$
  for all $j := 1$ to $\hat{p} - 1$ **do**
    $W := W + W_F^{c_{\mu_j}}(v|X_{\hat{p}})$ (equations (4.36)-(4.45))
  end for
  if $W > W^*$ **then**
    $Y := \{v\}$
    $W^* := W$
  else if $W = W^*$ **then**
    $Y := Y \cup \{v\}$
  end if
end for
if $Y = \emptyset$ **then**
  $Y := \{\hat{r}\}$
end if

Let us consider our example with the follower's location in vertex $v = 6'$, i.e. $k = 1$. We get $a_{c_{\mu_1}} = 2 - 2/2 = 1$, $b_{c_{\mu_1}} = (3 + 2)/2 = 2.5$, $a_{c_{\mu_2}} = 0$, $b_{c_{\mu_2}} = 5 - 3/2 = 3.5$, $a_{c_{\mu_3}} = 0$, $b_{c_{\mu_3}} = 0$, $a_{c_{\mu_4}} = 0$, $b_{c_{\mu_4}} = 0$ and $W_F(6'|X_{\hat{p}}) = 15.5$. Analogously, one computes the follower's market share when locating in vertex 1, 2, 3, 8 or 9. By comparison of these values we find vertex $6'$ to be the $(1|X_5)$-medianoid.

We may now conclude this section with the main result in Theorem 4.2.3.

Theorem 4.2.3. *The problem of finding a discrete, binary $(1|p)$-centroid of a tree network $N = (V, E, \lambda)$ with vertex and edge demand can be solved with time complexity $\mathcal{O}(n^2 p \log \xi(N))$.*

Proof. We apply bisection on the interval $[0, \xi(N)]$ to find the smallest value $\hat{\xi}$ such that a $\hat{\xi}$-bounding set with at most p elements exists (cf. Spoerhase and

Wirth, 2009c). The termination criterion described in the proof of Theorem 4.2.2 can be adapted to the case at hand. □

4.3 Conclusion

In this chapter we have analyzed a classical sequential location problem on networks, the $(r|p)$-centroid problem under a binary choice rule. Differing from the majority of previous publications, we have considered networks with both, vertex and edge demand. The corresponding follower problem is known to be NP-hard even on networks with edge demand only (Dasci et al., 2002). We have proven that this remains true for the $(r|p)$-centroid problem. On tree networks, however, an efficient algorithm to determine a $(1|p)$-centroid has been derived, after having restricted the location sites to the vertex set of the underlying network (discrete problem class). Bilevel programming models have been presented for the discrete $(r|p)$-centroid problem on general networks with vertex and edge demand. Computational results based on these models indicate, that heuristics will need to approximate the follower's response to a given set of leader's locations. An upper bound on the follower's optimal market share has previously been introduced for the case of vertex demand only. We have shown that this bound can be adapted to the case of vertex and edge demand. Future research topics (cf. also Section 4.1.4) include the application of non-uniform demand densities or different types of choice rules, as, for example, proportional or partially binary choice rules.

Chapter 5

Competitive Location and Pricing with Random Utilities

In this chapter we analyze the effect of including price competition into the discrete $(r|X_p)$-medianoid problem, sometimes also referred to as the *maximum capture problem* (MAXCAP, ReVelle, 1986), as presented in Kress and Pesch (2012a). We restrict our attention to the case of vertex demand only and contribute to the literature by extending two major publications:

- Benati and Hansen (2002) (cf. also Benati, 2000) introduce the maximum capture problem on networks with random utilities. Probabilistic customer behavior is modeled by applying the multinomial logit approach. In their definition of the deterministic utility component, the authors focus on incorporating effects of distances from customers to facility locations.
- Serra and ReVelle (1999) consider the maximum capture problem on networks under a binary choice rule, where players are allowed to compete in prices after having chosen locations.

As proposed by Benati and Hansen (2002), it is our aim to apply the idea of competition in prices by Serra and ReVelle (1999) to the maximum capture problem on networks with random utilities in order to "improve the realism of the model". In following the authors, we select the multinomial logit approach to model the decision process of utility maximizing customers. As a consequence, this chapter also contributes to numerical apprcaches for computing equilibrium prices under multinomial logit demand and is therefore very closely related to Morrow (2008); Morrow and Skerlos (2011). Related models can be found in the field of product positioning, cf. Choi et al. (1990) and Rhim and Cooper (2005).

This chapter proceeds as follows. First, a detailed problem formulation is given in Section 5.1 with results concerning the existence of price equilibria and the computational complexity in Sections 5.1.1 and 5.1.2, respectively. In Section 5.2 we show that, given the locations of all facilities, standard numerical approaches can be combined to reliably and quickly determine local price equilibria. Finally, in Section 5.3, we show that price competition may actually affect optimal locations of facilities, and we provide first insights into the performance of heuristic algorithms for finding solutions to the location problem. The chapter closes with a conclusion in Section 5.4.

5.1 Problem Formulation

Consider a network $N = (V, E, \lambda)$ with vertex demand only. A firm L (leader) acts as a monopolist with multiple facilities in this spatial market. L's facilities are located at $p > 0$ distinct vertices $X_p \subseteq V$ of the network. A competitor F (follower) wants to enter the market with an a priori fixed number of facilities $r > 0$.[1] F's potential facility sites are restricted to the vertex set of the network. Hence, F solves a discrete $(r|X_p)$-medianoid problem. At most two facilities, one of the leader's and one of the follower's facilities, may be located at each vertex. The players are profit maximizing and sell a single homogeneous product. As, in most markets, the choice of location is usually less flexible than the choice of prices, we assume that simultaneous price competition occurs after F's location decisions have been made. Thus, the game under consideration is composed of multiple stages (see Figure 5.1, cf. also Rhim and Cooper, 2005). In the first stage, F decides

1 An alternative model does not fix the number of facilities in advance, but incorporates fixed setup costs $f_j \in \mathbb{R}_0^+$ for (potential) facilities j; cf. Benati (2003) for a version of this problem without (explicit) price competition.

on the locations $Y_r \subseteq V$ of the facilities. In the second stage, both players simultaneously decide on a (mill) price for the product. This stage – as characterized by Eiselt et al. (1993) – is a noncooperative game with the strategies being prices and payoffs being profits. A solution to this stage is a pure strategy Nash equilibrium in prices, given that such an equilibrium exists. After the prices have been set, (homogenous) customers accommodate their demand and market shares are established.

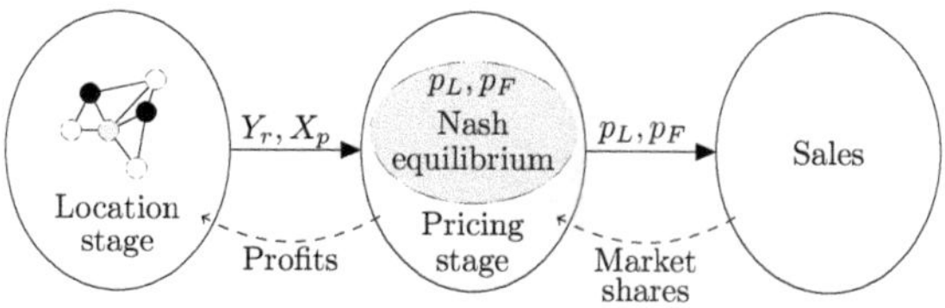

Fig. 5.1: Stages of the game

The utility u_{ij}^q of a customer located in vertex $i \in V$ from patronizing a facility located in vertex $j \in V$ and belonging to player $q \in \{L, F\}$ is composed of a deterministic component v_{ij}^q and a stochastic component ϵ_{ij}^q, the latter being related to unobservable, utility affecting factors:

$$u_{ij}^q = v_{ij}^q + \epsilon_{ij}^q. \tag{5.1}$$

Based on Benati and Hansen (2002), we define

$$v_{ij}^q := a_j^q - \alpha d_{ij} - \beta p_q, \tag{5.2}$$

where

- $a_j^q \in \mathbb{R}$ is the player-specific (index q) average quality level associated with a facility located in vertex j (related to opening hours, size, etc.),
- $\alpha \geq 0$ is a scaling parameter for distance ("coefficient of spatial friction", Benati and Hansen, 2002),
- $\beta > 0$ is a sensitivity parameter for price,
- p_q is the unit mill price charged at all of player q's facilities.

In the remainder of the chapter it will always become clear from the context, whether we refer to a leader's or a follower's facility. Hence, when not really needed, we will drop the player index in the following for the sake of simplicity and brevity, and denote the set of facilities (choice set) by simply writing $X_p \cup Y_r$. Following the multinomial logit approach as described in Chapter 1.1.3 and defining $s := 1/\sigma$ (assuming $\sigma, s > 0$) for the sake of notational convenience, the probability pr_{ij} that a customer located in vertex $i \in V$ chooses facility $j \in X_p \cup Y_r$ is:

$$pr_{ij} = \frac{e^{sv_{ij}}}{\sum\limits_{k \in X_p} e^{sv_{ik}} + \sum\limits_{k \in Y_r} e^{sv_{ik}}} \quad \forall i \in V, j \in X_p \cup Y_r. \tag{5.3}$$

As pointed out by Choi et al. (1990), the model lacks price elasticity, because every customer will choose a facility regardless of prices. Hence, as frequently done in the literature, an artificial facility ("no purchase" option) with a corresponding deterministic utility component of zero is included, so that we get

$$pr_{ij} = \frac{e^{sv_{ij}}}{\sum\limits_{k \in X_p} e^{sv_{ik}} + \sum\limits_{k \in Y_r} e^{sv_{ik}} + 1} \quad \forall i \in V, j \in X_p \cup Y_r. \tag{5.4}$$

Let $q \in \{L, F\}$ and define

$$Z := \begin{cases} X_p & \text{if } q = L, \\ Y_r & \text{if } q = F. \end{cases} \tag{5.5}$$

Moreover, let c_q be the cost of producing one unit of the product at one of player q's facilities. Then, given Y_r, p_F, X_p and p_L, the (expected) profit Π_q of player q is as follows:

$$\Pi_q = (p_q - c_q) \sum_{i \in V} \sum_{j \in Z} \pi(i) pr_{ij}. \tag{5.6}$$

We define the binary variables

$$y_j^F := \begin{cases} 1 & \text{if F locates a facility in vertex j,} \\ 0 & \text{else,} \end{cases} \quad \forall j \in V.$$

Then a mathematical programming formulation for the problem under consideration is as follows:

$$\max_{p_F, \boldsymbol{y}^F} \quad \Pi_F = (p_F - c_F) \sum_{i \in V} \sum_{j \in V} \pi(i) y_j^F \cdot \frac{e^{s(a_j - \alpha d_{ij} - \beta p_F)}}{\sum_{k \in X_p} e^{s(a_k - \alpha d_{ik} - \beta p_L)} + \sum_{k \in V} y_k^F e^{s(a_k - \alpha d_{ik} - \beta p_F)} + 1}, \tag{5.7}$$

$$\text{subject to} \quad p_L \in \operatorname*{argmax}_{p_L} \Pi_L = (p_L - c_L) \sum_{i \in V} \sum_{j \in X_p} \pi(i) \frac{e^{s(a_j - \alpha d_{ij} - \beta p_L)}}{\sum_{k \in X_p} e^{s(a_k - \alpha d_{ik} - \beta p_L)} + \sum_{k \in V} y_k^F e^{s(a_k - \alpha d_{ik} - \beta p_F)} + 1}, \tag{5.8}$$

$$p_F \in \operatorname*{argmax}_{p_F} \Pi_F = (p_F - c_F) \sum_{i \in V} \sum_{j \in V} \pi(i) y_j^F \frac{e^{s(a_j - \alpha d_{ij} - \beta p_F)}}{\sum_{k \in X_p} e^{s(a_k - \alpha d_{ik} - \beta p_L)} + \sum_{k \in V} y_k^F e^{s(a_k - \alpha d_{ik} - \beta p_F)} + 1}, \tag{5.9}$$

$$\sum_{j \in V} y_j^F = r, \tag{5.10}$$

$$p_L, p_F \geq 0, \tag{5.11}$$

$$y_j^F \in \{0, 1\} \qquad \forall j \in V. \tag{5.12}$$

The objective function (5.7) corresponds to the follower's (expected) profit maximization. Conditions (5.8) and (5.9) enforce a Nash equilibrium in prices, given that such an equilibrium exists. Restriction (5.10) guarantees that exactly r facility locations are selected. Conditions (5.11) and (5.12) define the domains of the variables. In the next section we will motivate the additional constraints

$$p_L, p_F \leq \bar{p} \tag{5.13}$$

that constitute an upper bound (parameter $\bar{p}$) on the prices charged by the players. We will refer to problem (5.7)–(5.13) as the *location-then-price game under a logit assumption* and denote it by LPL.

Throughout the remainder of this chapter, we will use index $q \in \{L, F\}$ to refer to the players of LPL. Moreover, given a player $q \in \{L, F\}$, we will refer to the opposing player by $\bar{q} := \{L, F\} \setminus q$.

5.1.1 Pricing Stage: Nash Equilibria and Local Equilibria

In this section, we provide a sufficient condition for the existence of a pure strategy Nash equilibrium in prices. To do so, we will require each payoff function to be concave in the corresponding player's price. Hence, for each player, a unique best response exists for each strategy of the opponent. Similar results are due to Choi et al. (1990) and Rhim and Cooper (2005). Note, however, that their proofs and discussions do not directly apply to the case $r, p > 1$. Furthermore, note that Morrow (2008); Morrow and Skerlos (2011) analyze very closely related games of finding Nash equilibria in prices. When analyzing the multinomial logit model, however, they do not allow for explicit (exogenous) upper bounds on the prices charged by the players, so that their results are not guaranteed to directly apply to LPL.

Lemma 5.1.1. *There exist finite upper bounds on the prices charged by the leader and the follower, i.e. $p_q < \infty$, $q \in \{L, F\}$.*

Proof. Define Z as in (5.5). We have

$$\lim_{p_q \to \infty} \Pi_q = \sum_{i \in V} \sum_{j \in Z} \pi(i) \cdot \lim_{p_q \to \infty} \frac{p_q - c_q}{pr_{ij}^{-1}}.$$

Using L'Hospital's rule for any $i \in V$, $j \in Z$, we get

$$\lim_{p_q \to \infty} \frac{p_q - c_q}{pr_{ij}^{-1}} = \lim_{p_q \to \infty} \frac{1}{\frac{\partial pr_{ij}^{-1}}{\partial p_q}} = \lim_{p_q \to \infty} \frac{1}{-\frac{1}{pr_{ij}^2} \frac{\partial pr_{ij}}{\partial p_q}} = \lim_{p_q \to \infty} \frac{1}{s\beta \left(\frac{1}{pr_{ij}} - \frac{\sum_{k \in Z} pr_{ik}}{pr_{ij}} \right)}$$

$$= \lim_{p_q\to\infty} \frac{1}{s\beta\left(\frac{1}{pr_{ij}} - \sum_{k\in Z} e^{s(a_k-a_j-\alpha(d_{ik}-d_{ij}))}\right)} = 0$$

because

$$\lim_{p_q\to\infty} pr_{ij} = \lim_{p_q\to\infty} \frac{e^{s(a_j-\alpha d_{ij}-\beta p_q)}}{\sum\limits_{k\in X_p} e^{s(a_k-\alpha d_{ik}-\beta p_L)} + \sum\limits_{k\in Y_r} e^{s(a_k-\alpha d_{ik}-\beta p_F)} + 1}$$

$$= \lim_{p_q\to\infty} \frac{1}{\sum\limits_{k\in Z} e^{s(a_k-a_j-\alpha(d_{ik}-d_{ij}))} + \frac{c}{e^{s(a_j-\alpha d_{ij}-\beta p_q)}}} = 0,$$

where

$$c := \begin{cases} \sum\limits_{k\in Y_r} e^{sv_{ik}} + 1 & \text{if } q = L, \\ \sum\limits_{k\in X_p} e^{sv_{ik}} + 1 & \text{if } q = F. \end{cases} \tag{5.14}$$

Thus, the players have no incentive to charge infinite prices. This proves the assertion. □

Observe that Π_q, $q \in \{L, F\}$, is negative for any $p_q < c_q$. Thus, it is reasonable to assume that prices are bounded below by the unit production costs, i.e. $c_q \leq p_q$. Additionally, motivated by Lemma 5.1.1, let $\bar{p} > \max\limits_{q\in\{L,F\}} c_q$ be some exogenous upper bound on the prices p_L and p_F. Then we may restrict player q's strategy space to the nonempty, compact and convex interval $[c_q, \bar{p}]$ (cf. conditions (5.13)).

Based on Theorem 1.1.1, a sufficient condition for the existence of a pure strategy Nash equilibrium in prices can easily be derived. To do so, we will require the payoff functions Π_q, $q \in \{L, F\}$, to be concave in p_q.

Theorem 5.1.1. *A sufficient condition for the existence of a pure strategy Nash equilibrium in prices is*

$$s\beta \leq \frac{2}{\bar{p} - c_q} \tag{5.15}$$

for $q \in \{L, F\}$.

Proof. It is easy to verify that the payoff functions (5.6) are continuous in $[c_L, \bar{p}] \times [c_F, \bar{p}]$. In the following we will derive sufficient conditions for the concavity of the payoff functions (5.6) in p_q.

Define Z as in (5.5) and consider an arbitrary vertex $i \in V$. Then, for any $j \in Z$,

$$pr_{ij} = \frac{e^{s(a_j - \alpha d_{ij} - \beta p_q)}}{\sum_{k \in X_p} e^{s(a_k - \alpha d_{ik} - \beta p_L)} + \sum_{k \in Y_r} e^{s(a_k - \alpha d_{ik} - \beta p_F)} - 1}.$$

It is easy to verify that

$$\frac{\partial pr_{ij}}{\partial p_q} = -s\beta pr_{ij} \left(1 - \frac{\sum_{k \in Z} e^{s(a_k - \alpha d_{ik} - \beta p_q)}}{\sum_{k \in X_p} e^{s(a_k - \alpha d_{ik} - \beta p_L)} + \sum_{k \in Y_r} e^{s(a_k - \alpha d_{ik} - \beta p_F)} + 1} \right) \tag{5.16}$$

$$= -s\beta pr_{ij} \left(1 - \sum_{k \in Z} pr_{ik} \right), \tag{5.17}$$

$$\frac{\partial^2 pr_{ij}}{\partial p_q^2} = s\beta pr_{ij} \sum_{k \in Z} \frac{\partial pr_{ik}}{\partial p_q} - s\beta \frac{\partial pr_{ij}}{\partial p_q} \left(1 - \sum_{k \in Z} pr_{ik} \right), \tag{5.18}$$

for all $i \in V$ and $j \in Z$, and

$$\frac{\partial \Pi_q}{\partial p_q} = \sum_{i \in V} \sum_{j \in Z} \pi(i) pr_{ij} + (p_q - c_q) \sum_{i \in V} \sum_{j \in Z} \pi(i) \cdot \frac{\partial pr_{ij}}{\partial p_q}, \tag{5.19}$$

$$\frac{\partial^2 \Pi_q}{\partial p_q^2} = 2 \sum_{i \in V} \sum_{j \in Z} \pi(i) \frac{\partial pr_{ij}}{\partial p_q} + (p_q - c_q) \sum_{i \in V} \sum_{j \in Z} \pi(i) \cdot \frac{\partial^2 pr_{ij}}{\partial p_q^2}. \tag{5.20}$$

Π_q, $q \in \{L, F\}$, is concave in p_q if

$$\frac{\partial^2 \Pi_q}{\partial p_q^2} \leq 0.$$

This is guaranteed if

$$2\frac{\partial pr_{ij}}{\partial p_q} + (p_q - c_q)s\beta pr_{ij}\sum_{k\in Z}\frac{\partial pr_{ik}}{\partial p_q} - (p_q - c_q)s\beta\frac{\partial pr_{ij}}{\partial p_q}\left(1 - \sum_{k\in Z} pr_{ik}\right) \leq 0$$

for all $i \in V$ and $j \in Z$. It is easy to see that $\partial pr_{ij}/\partial p_q < 0$ for all $i \in V$ and $j \in Z$. Thus,

$$(p_q - c_q)s\beta pr_{ij}\sum_{k\in Z}\frac{\partial pr_{ik}}{\partial p_q} \leq 0, \tag{5.21}$$

for all $i \in V$ and $j \in Z$, so that it is sufficient to require

$$2\frac{\partial pr_{ij}}{\partial p_q} - (p_q - c_q)s\beta\frac{\partial pr_{ij}}{\partial p_q}\left(1 - \sum_{k\in Z} pr_{ik}\right) \leq 0,$$

or, equivalently,

$$2 - (p_q - c_q)s\beta\left(1 - \sum_{k\in Z} pr_{ik}\right) \geq 0,$$

for all $i \in V$ and $j \in Z$. Given (5.15), the latter inequality holds for all $i \in V$ and $j \in Z$, since

$$0 < \left(1 - \sum_{k\in Z} pr_{ik}\right) < 1,$$

for all $i \in V$. This proves the claim. □

The derivation of a less restrictive existence result, for example by providing a better upper bound than the one used in (5.21), is left for future research. Similarly, apart from the rather trivial uniqueness result of Section 5.1.2, the "complexity of the demand function prohibits the derivation of a [more general] global uniqueness condition" (Choi et al., 1990), for example by requiring strict diagonal dominance of the Jacobian of the first order conditions for profit maximization (Gabay and Moulin, 1980). Hence, when conditions (5.15) do not hold, we must rely on local conditions for optimality of prices (Morrow, 2008; Morrow and Skerlos, 2011):

Definition 5.1.1 (Morrow (2008)). A price vector $\boldsymbol{p} = (p_L, p_F) \in [c_L, \bar{p}] \times [c_F, \bar{p}]$ is called a local (global) price equilibrium, if element p_q is a local (global) maximizer of $\Pi_q(p_q, \hat{p}_{\bar{q}})$ for each $q \in \{L, F\}$, where $\hat{p}_{\bar{q}}$ denotes a fixed price of player $\bar{q}$.

Note that any global price equilibrium is a pure strategy Nash equilibrium in prices. Additionally observe that, when conditions (5.15) hold, any local price equilibrium is a global price equilibrium.

5.1.2 Computational Complexity

In this subsection we will show that LPL is NP-hard.

Lemma 5.1.2. *Let*

$$s\beta \leq \frac{1}{\bar{p} - c_q} \tag{5.22}$$

for $q \in \{L, F\}$. *Then there exists a unique pure strategy Nash equilibrium in prices with* $p_L = p_F = \bar{p}$ *for all feasible location settings.*

Proof. Let $q \in \{L, F\}$ and Z as defined in (5.5). We will show that Π_q is strictly monotonic increasing in p_q on the interval $I = [c_q, \bar{p}]$ if (5.22) holds. This will prove the claim.

Π_q is strictly monotonic increasing on I if

$$\frac{\partial \Pi_q}{\partial p_q} = \sum_{i \in V} \sum_{j \in Z} \pi(i) pr_{ij} + (p_q - c_q) \sum_{i \in V} \sum_{j \in Z} \pi(i) \cdot \frac{\partial pr_{ij}}{\partial p_q} > 0$$

on I. It is sufficient to require

$$pr_{ij} + (p_q - c_q) \frac{\partial pr_{ij}}{\partial p_q} = pr_{ij} - (p_q - c_q) s\beta pr_{ij} \left(1 - \sum_{k \in Z} pr_{ik}\right) > 0$$

or, equivalently,

$$1 - (p_q - c_q)s\beta\left(1 - \sum_{k \in Z} pr_{ik}\right) > 0$$

for all $i \in V$ and $j \in Z$. It is easy to see that this condition holds if we assume (5.22) to hold. □

The maximum capture problem with random utilities (Benati, 2000; Benati and Hansen, 2002) (MAXCAP-R) is closely related to LPL. The former problem considers stages 1 (location) and 3 (sales) of Figure 5.1 only and uses $v'_{ij} := a'_j - \alpha' d_{ij}$ instead of (5.2).[2] A no purchase option is not considered in MAXCAP-R, i.e. we have (5.3) instead of (5.4). Furthermore, $c_L = c_F = 0$. (5.7)–(5.13) "reduces" to:[3]

$$\max_{\boldsymbol{y^F}} \quad \Pi'_F(\boldsymbol{y^F}) = \sum_{i \in V}\sum_{j \in V} \pi(i) y_j^F \cdot \frac{e^{s(a'_j - \alpha' d_{ij})}}{\sum_{k \in X_p} e^{s(a'_k - \alpha' d_{ik})} + \sum_{k \in V} y_k^F e^{s(a'_k - \alpha' d_{ik})}} \tag{5.23}$$

subject to $\quad (5.10), (5.12).$

Note that MAXCAP-R can not directly be interpreted to be a special case of LPL even if LPL's no purchase option is dropped, since we assume $\beta > 0$ and $\bar{p} > \max_{q \in \{L,F\}} c_q$.

Theorem 5.1.2 (Benati (2000); Benati and Hansen (2002)). *MAXCAP-R is NP-hard.*

The adaption of the proof of Theorem 5.1.2 to the MAXCAP-R with an additional no purchase option (MAXCAP-RNP) is relatively straight forward.[4] In the latter problem, the probability that a customer located in vertex $i \in V$ chooses facility $j \in X_p \cup Y_r$ is given by

2 Note that the player index has been dropped for the sake of simplicity, cf. already Section 5.1.

3 Note that, differing from Benati (2000), we allow co-location of the players.

4 While the following statements are essentially the same as in Benati (2000), modifications are needed with respect to the transformation of Dominating Set instances and the proofs of Lemmas 5.1.3 and 5.1.4.

$$pr'_{ij} = \frac{e^{s(a'_j - \alpha' d_{ij})}}{\sum\limits_{k \in X_p} e^{s(a'_k - \alpha' d_{ik})} + \sum\limits_{k \in Y_r} e^{s(a'_k - \alpha' d_{ik})} + 1}.$$

Given a set Y_r of follower locations, we define $z_i(Y_r) := \sum_{j \in Y_r} \pi(i) pr'_{ij}$.

The proof is based on a reduction of the NP-hard *Dominating Set* (DS) problem (Garey and Johnson, 1979): Given a network $N = (V, E, \lambda)$, $|V| = n$, with $\lambda(uv) = 1$ for all $[u, v] \in E$ and a positive integer $r \leq n$, is there a set $V' \subseteq V$ with $|V'| \leq r$ and $D(i, V') \leq 1$ for all $i \in V$?

Given an instance I_{DS} of DS, we construct an instance I_M of MAXCAP-RNP as follows. Set $\pi(v) = 1$ for every vertex $v \in V$ and add a vertex z with $\pi(z) = 0$. Denote the augmented vertex set by V'. Connect z to every vertex $v \in V$ by an edge $[z, v]$ of length $\lambda(zv) = 2$. Set $p = 1$ and locate the leader's facility in z. Moreover, choose $a'_j = 1.5\alpha'$ (more specifically $a_j^{F'}$) for all potential follower facility sites j and $a'_z = 1.5\alpha'$ (more specifically $a_z^{L'}$) for the leader facility in z. Set $s = 1$. We will prove (in analogy to Benati, 2000) that there exists a value $\alpha' > 0$ such that any optimal solution to I_M provides a dominating set of I_{DS} if one exists.

Given a solution Y_r to I_M, define $V_0 := \{v \in V | D(v, Y_r) = 0\}$, $V_1 := \{v \in V | D(v, Y_r) = 1\}$ and $V_2 := \{v \in V | D(v, Y_r) \geq 2\}$. Note that Y_r may contain vertex z while this is not the case for the sets V_0 to V_2. V_0 and V_1 are referred to as dominated sets of V, V_2 is the non-dominated set of V. The elements are called dominated and non-dominated vertices, respectively. Observe that $z_z(Y_r) = 0$.

Lemma 5.1.3. *Let $i \in V_2$. Then, for every $\epsilon_1 > 0$, there exists a value α'_1, such that for every $\alpha' \geq \alpha'_1$ we have $z_i(Y_r) \leq \epsilon_1$.*

Proof. It is easy to see that $\sum_{j \in Y_r} e^{a'_j - \alpha' d_{ij}} \leq re^{-0.5\alpha'}$. Therefore,

$$z_i(Y_r) \leq \frac{re^{-0.5\alpha'}}{re^{-0.5\alpha'} + e^{-0.5\alpha'} + 1} = \frac{r}{r + 1 + e^{0.5\alpha'}}.$$

Now $\lim\limits_{\alpha' \to \infty} \frac{r}{r+1+e^{0.5\alpha'}} = 0$ which proves the claim. □

Lemma 5.1.4. *Let $i \in V_0 \cup V_1$. Then, for every $\epsilon_2 > 0$, there exists a value α'_2, such that for every $\alpha' \geq \alpha'_2$ we have $z_i(Y_r) \geq 1 - \epsilon_2$.*

Proof. It is easy to see that $\sum_{j\in Y_r} e^{a'_j-\alpha' d_{ij}} \geq e^{0.5\alpha'}$. Therefore,

$$z_i(Y_r) \geq \frac{e^{0.5\alpha'}}{e^{0.5\alpha'}+e^{-0.5\alpha'}+1} = \frac{1}{1+e^{-\alpha'}+e^{-0.5\alpha'}} = 1 - \frac{e^{-\alpha'}+e^{-0.5\alpha'}}{1+e^{-\alpha'}+e^{-0.5\alpha'}}.$$

Now $\lim\limits_{\alpha'\to\infty}\left(1-\frac{e^{-\alpha'}+e^{-0.5\alpha'}}{1+e^{-\alpha'}+e^{-0.5\alpha'}}\right) = 1$ which proves the claim. □

Lemma 5.1.5 (Benati (2000)). *There exists a finite value $\hat{\alpha}'$ such that any optimal solution to I_M dominates the maximum number of vertices of V.*

Proof. Let Y_r^1 and Y_r^2 be two feasible solutions to I_M with $|V_0 \cup V_1| = \tau$ and $|V_0 \cup V_1| = \kappa$, respectively. That is, Y_r^1 dominates τ vertices and Y_r^2 dominates κ vertices of V. Assume $\tau > \kappa$. Then, from Lemma 5.1.4 we get $\sum_{i\in V'} z_i(Y_r^1) \geq \tau(1-\epsilon_2)$, where $\epsilon_2 := \frac{e^{-\alpha'}+e^{-0.5\alpha'}}{1+e^{-\alpha'}+e^{-0.5\alpha'}}$. Similarly, from Lemma 5.1.3 we get $\sum_{i\in V'} z_i(Y_r^2) \leq \kappa + (n-\kappa)\epsilon_1$, where $\epsilon_1 := \frac{r}{r+1+e^{0.5\alpha'}}$.

Define $\epsilon := \max\{\epsilon_1, \epsilon_2\}$. We want to guarantee that $\sum_{i\in V'} z_i(Y_r^1) > \sum_{i\in V'} z_i(Y_r^2)$. A sufficient condition is $\tau(1-\epsilon) > \kappa+(n-\kappa)\epsilon$, or, equivalently, $\tau-\kappa > (n-\kappa+\tau)\epsilon$. It is easy to see that the latter statement is true if $\epsilon < 1/(2n)$, because $\tau-\kappa \geq 1$ and $(n-\kappa+\tau) \leq 2n$. Since we can make ϵ arbitrarily small by increasing α', we have proven the claim. □

It is easy to confirm that $\epsilon < 1/(2n)$ (as required in the proof of Lemma 5.1.5) is true if $\alpha' > 2\ln(4rn)$. Therefore, there exists a polynomially bounded, finite value $\hat{\alpha}'$ which guarantees that any optimal solution to I_M provides a dominating set of I_{DS} if one exists. The fact that MAXCAP-RNP is NP-hard follows readily:

Theorem 5.1.3. *MAXCAP-RNP is NP-hard.*

We conclude:

Theorem 5.1.4. *LPL is NP-hard.*

Proof. Consider an arbitrary instance I_M of MAXCAP-RNP. Now construct an instance I_{LPL} of LPL on the same network and with the same number and predefined locations of leader and follower facilities, by setting $c_L = c_F = 0$ and choosing arbitrary values $\bar{p} > 0$ and $\beta > 0$ such that conditions (5.22) hold. I_{LPL} reduces to a pure location game with parametric prices $p_L = p_F = \bar{p}$ (Lemma 5.1.2). Set $\alpha = \alpha'$ and $a_j = a'_j + \beta\bar{p}$ for all of the players' potential facility locations j.

It is easy to see that any optimal solution of I_{LPL} is optimal for I_M as well. Observe that the objective function values differ by a factor of $\bar{p}$ as chosen above. Thus, we have shown that, given an instance of the NP-hard MAXCAP-RNP, there exists a polynomial transformation to an instance of LPL, which, in turn, proves that LPL is NP-hard. □

5.2 Pricing Stage: Algorithmic Approaches

In this section we will analyze numerical approaches to computing local price equilibria. Since the players are profit maximizers, the problem of finding an equilibrium corresponds to finding a solution to the system

$$\begin{pmatrix} \frac{\partial \Pi_L}{\partial p_L} \\ \frac{\partial \Pi_F}{\partial p_F} \end{pmatrix} = \begin{pmatrix} 0 \\ 0 \end{pmatrix} \tag{5.24}$$

of nonlinear equations (first order conditions), if a solution exists, and verifying that the solution locally maximizes the payoff function of each player in the player's price (second order conditions). It is easy to derive the second order condition for profit maximization of player $q \in \{L, F\}$, i.e.

$$\begin{aligned} \frac{\partial^2 \Pi_q}{\partial p_q^2} &= -2s\beta \sum_{i \in V} \pi(i)(Z_{iq} - Z_{iq}^2) \\ &\quad + (p_q - c_q)s^2\beta^2 \sum_{i \in V} \pi(i)(2Z_{iq}^3 - 3Z_{iq}^2 + Z_{iq}) < 0, \end{aligned} \tag{5.25}$$

where Z is defined as in (5.5) and $Z_{iq} := \sum_{k \in Z} pr_{ik}$ for a given $i \in V$, by applying results of Section 5.1.1 (see, in particular, (5.16)–(5.20)).

Observe, however, that, due to the restriction of the players' strategy spaces by upper and lower bounds (cf. Section 5.1.1), (5.24) will not necessarily have a solution, even if a local equilibrium in prices exists. Figure

5.2 depicts an example instance.[5] The curves divide the plane into areas of positive and negative partial derivatives of the players' profit functions. The unique Nash equilibrium is $p_L^N = \bar{p} = 100$ ($\partial\Pi_L/\partial p_L > 0$), $p_F^N = 85.87$ ($\partial\Pi_F/\partial p_F = 0$). We will refer to equilibria of this type as *degenerate*. Algorithmic approaches will have to suitably address the potential existence of such degenerate equilibria.

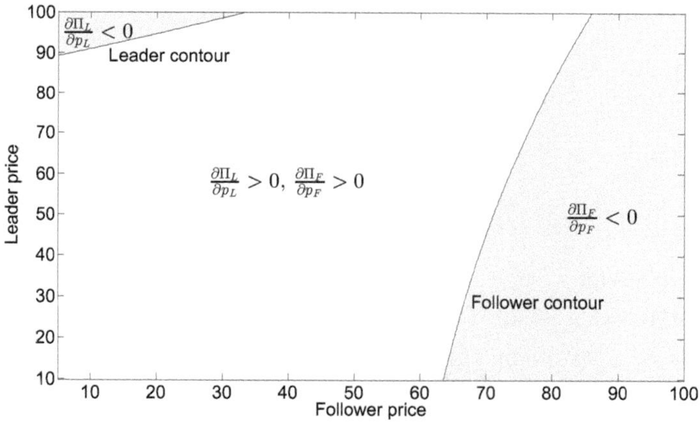

Fig. 5.2: Nash equilibrium with $\partial\Pi_L/\partial p_L > 0$

As in the study of Morrow and Skerlos (2011), natural (numerical) candidates to solving (5.24) include Newton's method (NM) and, due to the existence of fixed-point equations representing (5.24), fixed-point iteration (FPI) approaches. However, as pointed out in Section 5.1.1, their results are not guaranteed to directly apply to LPL. Hence, the following subsections are devoted to analyzing numerical methods for the case under consideration in this chapter and designing a reliable algorithm to determine local price equilibria. As we will see in Section 5.2.4, spatial friction has a major impact on the convergence behavior of numerical approaches to finding local price equilibria and on the existence of such equilibria.

Choi et al. (1990) and Rhim and Cooper (2005), among others, apply a variational inequality approach to compute Nash equilibria. In the context

5 Instance details: $n = 100$, $r = p = 5$, $\lambda(uv) \in [1, 50]$ for all $[u, v] \in E$, $\pi(v) \in [0, 100]$ for all $v \in V$, $a_j \in [0, 50]$ for all $j \in X_p \cup Y_r$, $\alpha = 0.287$, $\beta = 0.02$, $\bar{p} = 100$, $c_L = 9.78$, $c_F = 5.04$, $s = 1$.

of LPL, this approach is based on the fact that, assuming the players' payoff functions to be pseudoconcave in the corresponding player's prices, a price vector (p_L^N, p_F^N) is a Nash equilibrium if and only if it is a solution to the variational inequality

$$\sum_{q\in\{L,F\}} \left(-\frac{\partial \Pi_q}{\partial p_q}(p_L^N, p_F^N) \cdot (p_q - p_q^N) \right) \geq 0 \, \forall (p_L, p_F) \in [c_L, \bar{p}] \times [c_F, \bar{p}] \quad (5.26)$$

(cf. Gabay and Moulin, 1980). The variational inequality (5.26) may be solved by solving a series of nonlinear optimization problems

$$\max_{(p_L,p_F)\in[c_L,\bar{p}]\times[c_F,\bar{p}]} \sum_{q\in\{L,F\}} \Pi_q(p_q, \hat{p}_{\bar{q}}^k), \quad (5.27)$$

where, in iteration k, $\hat{p}_{\bar{q}}^k$ corresponds to a price that has been fixed at the previous iteration's value (Choi et al., 1990). In the case of convergence, this results in a "tatonnement" process towards a Nash equilibrium. A major drawback of this approach is, among others, the necessity of having to solve a series of optimization problems. In each of these solution processes, divergence issues may arise. Hence, we will not consider this approach as a reliable candidate to solving (5.24).

5.2.1 Fixed-Point Iteration

One easily derives a simple fixed-point formulation of the first order conditions for profit maximization by substituting (5.17) and (5.19) into (5.24) and applying some straightforward algebraic operations:

$$\begin{pmatrix} p_L \\ p_F \end{pmatrix} = \begin{pmatrix} c_L \\ c_F \end{pmatrix} + \frac{1}{s\beta} \begin{pmatrix} \frac{\sum_{i\in V} \pi(i) Z_{iL}}{\sum_{i\in V} \pi(i)(Z_{iL} - Z_{iL}^2)} \\ \frac{\sum_{i\in V} \pi(i) Z_{iF}}{\sum_{i\in V} \pi(i)(Z_{iF} - Z_{iF}^2)} \end{pmatrix}. \quad (5.28)$$

Hence, as pointed out above, FPI is a natural candidate to computing local price equilibria. We assume the reader to be familiar with the basics of FPI and refer to Quarteroni et al. (2000) for an introduction.

In analogy to Morrow and Skerlos (2011) we will refer to (5.28) as the *markup equation.*[6] This name is motivated by the fact that, on the right hand side of (5.28), a nonnegative value (markup) is added to the cost c_q of player $q \in \{L, F\}$. As in the framework of Morrow and Skerlos (2011), the (corresponding) markup equation is no reliable method for determining local price equilibria. This fact is illustrated in Figure 5.3.

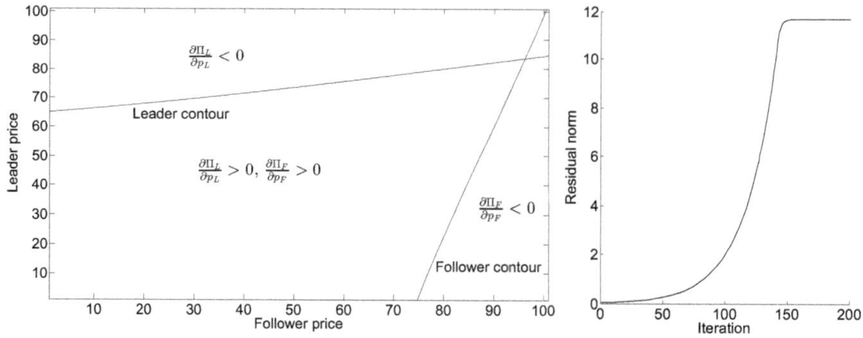

(a) Unique Nash equilibrium *(b) Convergence behavior*

Fig. 5.3: Diverging fixed-point iteration

The figure depicts a contour plot of the leader's and follower's partial profit derivatives of a given problem instance[7] (Figure 5.3a) as well as a plot of the corresponding convergence behavior (residual norm, i.e. the Euclidean norm of the vector pointing from the price pair of a specific iteration to the price pair of the previous iteration, over the number of iterations) of the FPI based on (5.28) (Figure 5.3b). The instance has a unique Nash equilibrium $p_L^N = 83.422$, $p_F^N = 96.196$ in prices (Figure 5.3a). However, although the FPI starts in point $(83.4, 96.2)$ very close to the Nash equilibrium, Figure 5.3b shows a rapid divergence behavior with the method ending up in circling between a set of (two) price pairs. We have frequently encountered this type of divergence behavior in computational tests and conclude that FPI

6 More specifically, (5.28) corresponds to the "η-markup equation" in Morrow and Skerlos (2011).

7 Instance details: $n = 100$, $r = 5$, $p = 3$, $\lambda(uv) \in [1, 50]$ for all $[u, v] \in E$, $\pi(v) \in [0, 100]$ for all $v \in V$, $a_j \in [0, 20]$ for all $j \in X_p \cup Y_r$, $\alpha = 0.704$, $\beta = 0.184$, $\bar{p} = 101$, $c_L = 1.06$, $c_F = 1.05$, $s = 0.1$. Observe that (5.15) holds for $q \in \{L, F\}$.

based on the markup equation (5.28) is (on its own) no reliable approach to determining local price equilibria in general.

As mentioned above, we will have to take account of the bounds on the players' prices when implementing FPI, i.e. we may not allow prices $p_q > \bar{p}$, $q \in \{L, F\}$. We do so by enforcing $p_q = \bar{p}$ when the corresponding markup becomes too large.[8] It is easy to see that, as a byproduct, this enables us to potentially find degenerate local price equilibria.

5.2.2 Newton's Method

Another well known technique for computing zeros of nonlinear systems is Newton's method (cf., for example, Kelley, 2003). In the context of LPL, a straight forward adaption of Newton's method directly solves (5.24) after having substituted (5.17) and (5.19). However, the profit derivatives vanish for large prices. This problem is found by Morrow and Skerlos (2011) as well: "The challenge is the tendency of the derivatives of the profits to vanish as prices become large".

Lemma 5.2.1. $\lim\limits_{p_q \to \infty} \frac{\partial \Pi_q}{\partial p_q} = 0$ *for* $q \in \{L, F\}$.

Proof. Making use of the results of Section 5.1.1 and defining Z as in (5.5), we get

$$\lim_{p_q \to \infty} \frac{\partial \Pi_q}{\partial p_q} = \lim_{p_q \to \infty} \left(\sum_{i \in V} \sum_{j \in Z} \pi(i) pr_{ij} + (p_q - c_q) \sum_{i \in V} \sum_{j \in Z} \pi(i) \cdot \frac{\partial pr_{ij}}{\partial p_q} \right)$$
$$= -s\beta \lim_{p_q \to \infty} (p_q - c_q) \cdot \sum_{i \in V} \sum_{j \in Z} \pi(i) pr_{ij} \left(1 - \sum_{k \in Z} pr_{ik} \right) = 0,$$

because, by applying L'Hospital's rule, we derive

8 Note, however, that divergence issues as in Figure 5.3 arise independently of the upper bound on the players' prices.

$$\lim_{p_q\to\infty} p_q pr_{ij} = \lim_{p_q\to\infty} \frac{p_q}{\sum_{k\in Z} e^{s(a_k-a_j-\alpha(d_{ik}-d_{ij}))} + \frac{c}{e^{s(a_j-\alpha d_{ij}-\beta p_q)}}}$$
$$= \lim_{p_q\to\infty} \frac{e^{s(a_j-\alpha d_{ij}-\beta p_q)}}{cs\beta} = 0$$

for any $i \in V$, $j \in Z$ and c as defined in (5.14). □

Hence, an implementation of NM may spuriously terminate at points that do not represent local price equilibria. Figure 5.4a illustrates this challenge based on an example instance.[9] Morrow and Skerlos (2011) therefore suggest using a more reliable implementation of Newton's method, based on rearranging the markup equations (5.28). In the context of LPL, we derive the nonlinear system

$$\begin{aligned}\mathbf{M}(\mathbf{p}) &:= \begin{pmatrix} M_L(\mathbf{p}) \\ M_F(\mathbf{p}) \end{pmatrix} \\ &:= \begin{pmatrix} p_L \\ p_F \end{pmatrix} - \begin{pmatrix} c_L \\ c_F \end{pmatrix} - \frac{1}{s\beta} \begin{pmatrix} \frac{\sum_{i\in V}\pi(i)Z_{iL}}{\sum_{i\in V}\pi(i)(Z_{iL}-Z_{iL}^2)} \\ \frac{\sum_{i\in V}\pi(i)Z_{iF}}{\sum_{i\in V}\pi(i)(Z_{iF}-Z_{iF}^2)} \end{pmatrix}.\end{aligned} \tag{5.29}$$

It is easy to see that any solution to $\mathbf{M}(\mathbf{p}) = \mathbf{0}$ solves (5.24) as well. Furthermore, observe that $\lim_{p_q\to\infty} M_q(\mathbf{p}) = \infty$ for $q \in \{L, F\}$. Figure 5.4b illustrates the characteristics of (5.29) for large prices with the example instance.

Our implementation of NM makes use of the analytic Jacobian matrix $\mathbf{J}(\mathbf{p})$ of (5.29). Let $q \in \{L, F\}$ and define Z as in (5.5). Then, for any $i \in V$ and $j \in Z$,

$$\frac{\partial pr_{ij}}{\partial p_{\bar{q}}} = s\beta pr_{ij} Z_{i\bar{q}},$$

and thus

9 Instance details: $n = 100$, $r = p = 5$, $\lambda(uv) \in [1, 50]$ for all $[u, v] \in E$, $\pi(v) \in [0, 100]$ for all $v \in V$, $a_j \in [0, 30]$ for all $j \in X_p \cup Y_r$, $\alpha = 1.982$, $\beta = 1.342$, $\bar{p} = 150$, $c_L = 6.442$, $c_F = 8.334$, $s = 1$.

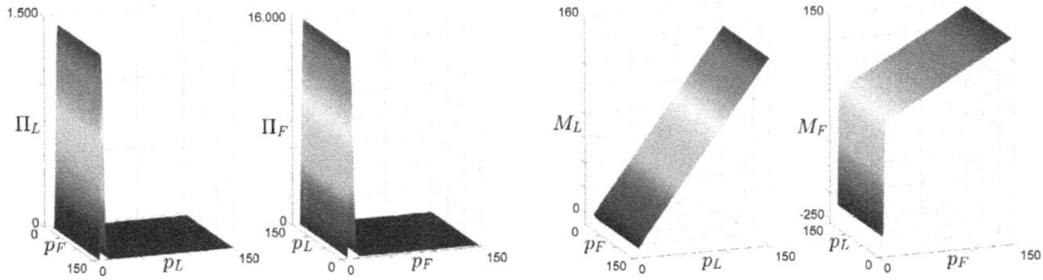

(a) *Profit functions* (b) *Markup functions*

Fig. 5.4: Comparison of profit and markup functions

$$\mathbf{J}(\mathbf{p}) := \begin{bmatrix} \frac{\partial M_L(\mathbf{p})}{\partial p_L} & \frac{\partial M_L(\mathbf{p})}{\partial p_F} \\ \frac{\partial M_F(\mathbf{p})}{\partial p_L} & \frac{\partial M_F(\mathbf{p})}{\partial p_F} \end{bmatrix}$$

with

$$\frac{\partial M_q}{\partial p_q} = \frac{2\left(\sum_{i\in V}\pi(i)(Z_{iq}-Z_{iq}^2)\right)^2 + \left(\sum_{i\in V}\pi(i)Z_{iq}\right)\left(\sum_{i\in V}\pi(i)(-Z_{iq}+3Z_{iq}^2-2Z_{iq}^3)\right)}{\left(\sum_{i\in V}\pi(i)(Z_{iq}-Z_{iq}^2)\right)^2},$$

$$\frac{\partial M_q}{\partial p_{\bar{q}}} = \frac{-\left(\sum_{i\in V}\pi(i)(Z_{iq}-Z_{iq}^2)\right)\left(\sum_{i\in V}\pi(i)Z_{iq}Z_{i\bar{q}}\right) + \left(\sum_{i\in V}\pi(i)Z_{iq}\right)\left(\sum_{i\in V}\pi(i)(Z_{iq}Z_{i\bar{q}}-2Z_{iq}^2Z_{i\bar{q}})\right)}{\left(\sum_{i\in V}\pi(i)(Z_{iq}-Z_{iq}^2)\right)^2}.$$

The dimension of $\mathbf{J}(\mathbf{p})$ enables us to easily calculate its inverse $\mathbf{J}^{-1}(\mathbf{p})$ to directly determine the (full) Newton step $\mathbf{p}_{k+1} - \mathbf{p}_k = -\mathbf{J}^{-1}\mathbf{M}$ in each iteration k. We apply a line search method, i.e. a standard global convergence strategy based on enforcing a (sufficient) decrease of $\|\mathbf{M}\|_2^2$, in every iteration. We refer the reader to Press et al. (2002, Chapter 9.7) for the details of our line search implementation (cf. also Kelley, 2003, Chapter 1).

For the example of Figure 5.3, our NM implementation converges to the unique Nash equilibrium within four iterations when started at (c_L, c_F). In general, however, applying NM may collide with the need to find degenerate price equilibria, as such equilibria do not solve (5.24) and $\mathbf{M}(\mathbf{p}) = \mathbf{0}$. In such cases, stepping towards a degenerate equilibrium may cause an increase of $\|\mathbf{M}\|_2^2$. Hence, while frequently outperforming FPI in cases like the one of

Figure 5.3 in our computational tests, NM applied to (5.29) is, on its own, no reliable method to find local price equilibria of general LPL instances with fixed sets of locations.

5.2.3 A Combined Algorithm

In the previous sections we have shown that neither the proposed FPI nor NM are, on their own, adequate methods for determining local price equilibria. However, the characteristics of these methods are quite different with respect to their potential of finding degenerate equilibria and their convergence behavior. Therefore, it is reasonable to develop a combined algorithm that exploits the strengths of both methods. Figure 5.5 depicts the flowchart of a combined algorithm (CA), that we will describe in detail in this section.

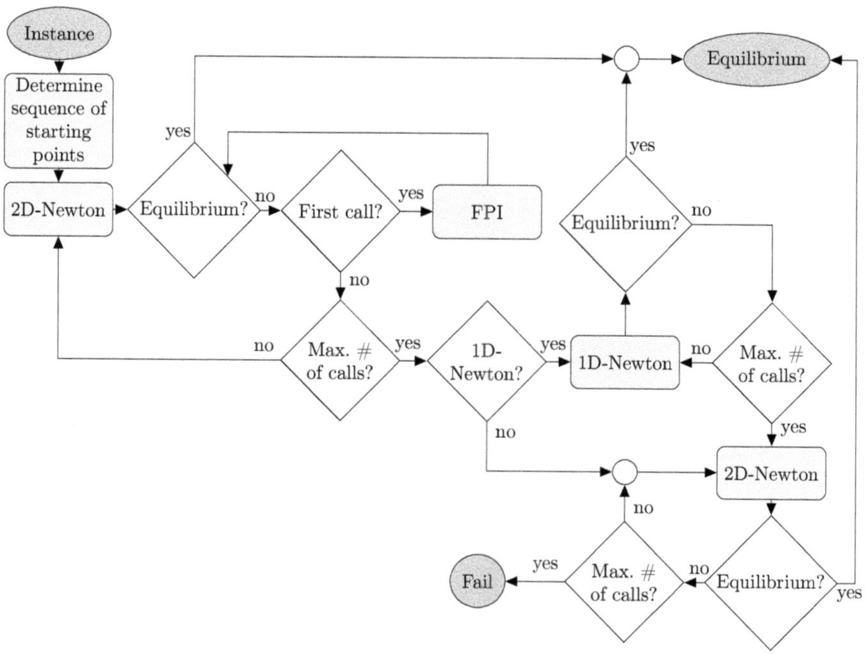

Fig. 5.5: Combined algorithm

The algorithm's parameter values are based on our computational experience and result in a reliable and fast convergence behavior (see Section 5.2.4).

CA starts with repeatedly calling NM (first "2D-Newton" block) with a sequence of eight starting price vectors. Each call of NM is limited to 230 iterations. The starting prices are chosen based on the relative "*strength*" of the players. We apply a very simple measure of player q's, $q \in \{L, F\}$, strength, i.e.

$$S_q := \begin{cases} \frac{1}{c_q}\left(\sum_{j\in Z} a_j - \alpha\frac{\sum_{j\in Z}\sum_{i\in V} d_{ij}}{n}\right) & \text{if } c_q \geq 1 \text{ and } c_{\bar{q}} \geq 1, \\ \sum_{j\in Z} a_j - \alpha\frac{\sum_{j\in Z}\sum_{i\in V} d_{ij}}{n} & \text{else,} \end{cases} \tag{5.30}$$

and call player q *dominant*, if $S_{\bar{q}} \geq 0$ and $S_q > 3S_{\bar{q}}$ or if $S_{\bar{q}} < 0$ and $S_q > 0.75S_{\bar{q}}$. CA's sequence of starting vectors is illustrated in Figure 5.6.

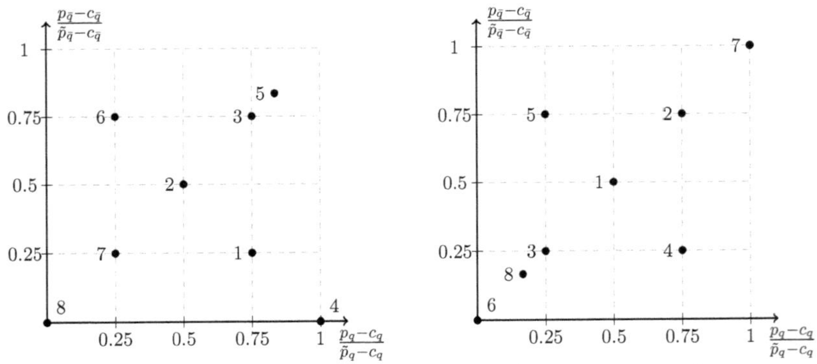

(a) Dominant player q *(b) No dominant player*

Fig. 5.6: Starting price vectors

To avoid starting NM in areas of low profits and small partial profit derivatives of both players (recall that $\lim_{p_q\to\infty}\Pi_q = 0$ and $\lim_{p_q\to\infty}\frac{\partial\Pi_q}{\partial p_q} = 0$ for $q \in \{L, F\}$), we determine price level $\tilde{p}_q$, $q \in \{L, F\}$, by applying Algorithm 5.2.1.

Algorithm 5.2.1 Determine $\tilde{p}_q$, $q \in \{L, F\}$

$ub_q := \bar{p}$, $\Delta_q := 0$, $p_q := \bar{p}$, $\tilde{p}_q := \bar{p}$, $q \in \{L, F\}$
$it := 1$, $go := true$, $pp := false$
while go **do**
    $\Delta_q := 0.1(ub_q - c_q)$, $q \in \{L, F\}$

```
    it := 1
    while go and it ≤ 10 do
        if Π_q(p_L, p_F) ≤ 10^-4, q ∈ {L, F} then
            if it ≠ 10 then
                p_q := p_q − Δ_q, q ∈ {L, F}
            end if
            if |p_q − c_q| ≤ 1 for any q ∈ {L, F} then
                go := false
            end if
            if it = 10 then
                ub_q := p_q, q ∈ {L, F}
            end if
        else
            go := false
            pp := true
        end if
        it := it + 1
    end while
end while
if p_q < p̄, q ∈ {L, F} and pp then
    p̃_q := p_q + Δ_q, q ∈ {L, F}
else
    p̃_q := p̄, q ∈ {L, F}
end if
```

If the first NM call does not converge to a local price equilibrium, we start FPI with the price vector that has been examined in the last iteration of the NM call, or, in case of convergence to a local minimum of a player's profit function, in point $(0.5(\bar{p}-c_L)+c_L, 0.5(\bar{p}-c_F)+c_F)$. This way we are able to exploit FPI's potential of finding degenerate equilibria. As before, we set the maximum number of FPI iterations to 230.

As mentioned above, we may have to abort a NM call due to the restriction of the players' prices by upper and lower bounds, for example if the line search does not find a step towards a local equilibrium in prices that reduces

$||\mathbf{M}||_2^2$. However, we may make use of such abortions, as they may point to the existence of degenerate price equilibria. Hence, if neither one of the calls of NM nor FPI have resulted in finding a local price equilibrium and if the calls indicate the existence of a degenerate equilibrium, we make use of this information by fixing the corresponding player's price to its bound and calling a standard one-dimensional Newton's method (cf., for example, Press et al., 2002, Chapter 9.7) to find a locally optimal reaction of the opposing player. Our implementation of the one-dimensional Newton's method ("1D-Newton" block) does not use any global convergence strategy. It is called at most fifteen times for each player. Each call is restricted to at most 100 iterations. In the first call, we set the starting price to the average price returned by the previous NM calls. Subsequently, we increase the corresponding player's unit production cost by a constant increment until we reach $\bar{p}$.

We have very rarely encountered instances, where a local price equilibrium could not be not found with the method described so far (see Section 5.2.4). To appropriately handle those instances, we have implemented another cycle of NM calls (second "2D-Newton" block) with another 100 potential starting price vectors, that are equally distributed over the domain of prices. Hence, in contrast to the prior NM calls, starting prices are independent of the players' and facilities' characteristics. Alternatively, if β is very large, we call an *alternating Newton* approach, i.e. a variant of the tatonnement approach by Choi et al. (1990). Starting in $(p_L, p_F) = (c_L, c_F)$, this approach alternately fixes the price of one player and makes a single call of the one-dimensional Newton's method to compute a locally optimal reaction of the opposing player that is then fixed in the subsequent iteration.

We close this subsection by describing CA's procedure of validating local price equilibria ("Equilibrium?"-decision blocks in Figure 5.5). Given a price vector $\mathbf{p} = (p_L, p_F)$ with both elements smaller than $\bar{p}$, we check (5.25) for both players $q \in \{L, F\}$, (5.24) and $\mathbf{M}(\mathbf{p}) = \mathbf{0}$. Similarly, if exactly one element p_q, $q \in \{L, F\}$, of $\mathbf{p}$ is at its upper bound, we check the opposing player $\bar{q}$'s second order condition (5.25), $\partial\Pi_{\bar{q}}/\partial p_{\bar{q}} = 0$, and $M_{\bar{q}}(\mathbf{p}) = 0$. For each element p_q, $q \in \{L, F\}$, of $\mathbf{p}$, however, that is at its upper bound, we check, if

- $\frac{\partial \Pi_q}{\partial p_q}(\mathbf{p}) > 0$, or
- $\frac{\partial \Pi_q}{\partial p_q}(\mathbf{p}) = 0$ and $\frac{\partial^2 \Pi_q}{\partial p_q^2}(\mathbf{p}) < 0$, cf. (5.25), or
- $\frac{\partial \Pi_q}{\partial p_q}(\mathbf{p}) = 0$, $\frac{\partial^2 \Pi_q}{\partial p_q^2}(\mathbf{p}) = 0$, cf. (5.25), and $\frac{\partial \Pi_q}{\partial p_q}(p_q - \epsilon, p_{\bar{q}}) > 0$ for a sufficiently small ϵ.

Additionally, if we have manually set a player's price to its upper bound, we check if the corresponding player's relative market share is reasonably large, i.e. if $(\sum_{i \in V} \pi(i) Z_{iq})/(\sum_{i \in V} \pi(i)) \geq 10^{-3}$.

5.2.4 Computational Analysis

In order to verify the stability and to analyze the performance of CA, we have conducted a series of computational experiments with the location stage being excluded from the numerical tests by randomly selecting the players' facility sites from the vertex set of the test networks. All test instances have been generated randomly with each parameter being drawn from a uniform distribution over a specific interval. The underlying networks are complete with edge length $\lambda(uv) \in [1, 50]$ for all $[u, v] \in E$ and customer demand $\pi(u) \in [0, 100]$ for all $u \in V$. Moreover, $c_q \in [1, 10]$, $q \in \{L, F\}$. If not stated otherwise, we fix s to one (as frequently done in the literature), we set $n = 100$, and we select a_j from the interval $[10, 50]$ for all $j \in Y_r \cup X_p$. In order to keep a somewhat realistic flavor, we additionally enforce the sum of the players' facility quality levels, $\sum_{j \in Z} a_j$ for $q \in \{L, F\}$ and Z as defined in (5.5), as well as the average quality levels, $(\sum_{j \in Z} a_j)/|Z|$, to differ by no more than a factor of five.

Sensitivity parameters for price and distance typically range from zero to about 4 in the literature (for similar ranges of the other parameters as above), see, for example, Benati (2000); Benati and Hansen (2002); Rhim and Cooper (2005); Thomadsen (2005). We have therefore generated five sets of LPL instances with $n = 100$ and 10,000 instances in each set. Table 5.1 presents the underlying ranges for the random generation of the corresponding parameters. Note that Theorem 5.1.1 holds for every instance of

Set 1. Hence, the players' payoff functions are concave and every local price equilibrium is a pure strategy Nash equilibrium in prices.

Tab. 5.1: Sets of test instances

	α	β	$\bar{p}$	s	$a_j \forall j \in X_P \cup Y_r$	r	p
Set 1	$[0.015, 0.4]$	$[0.015, 0.2]$	100	0.1	$[0, 20]$	$[1, 5]$	$[1, 5]$
Set 2	$[0.015, 0.5]$	$[0.015, 0.5]$	150	1	$[10, 50]$	$[1, 5]$	$[1, 5]$
Set 3	$[0.015, 1]$	$[0.015, 1]$	150	1	$[10, 50]$	$[1, 5]$	$[1, 5]$
Set 4	$[0.015, 2]$	$[0.015, 2]$	150	1	$[10, 50]$	$[1, 5]$	$[1, 5]$
Set 5	$[0.015, 4]$	$[0.015, 4]$	150	1	$[10, 50]$	$[1, 5]$	$[1, 5]$

Figure 5.7 provides results on CA's convergence behavior for the test sets.

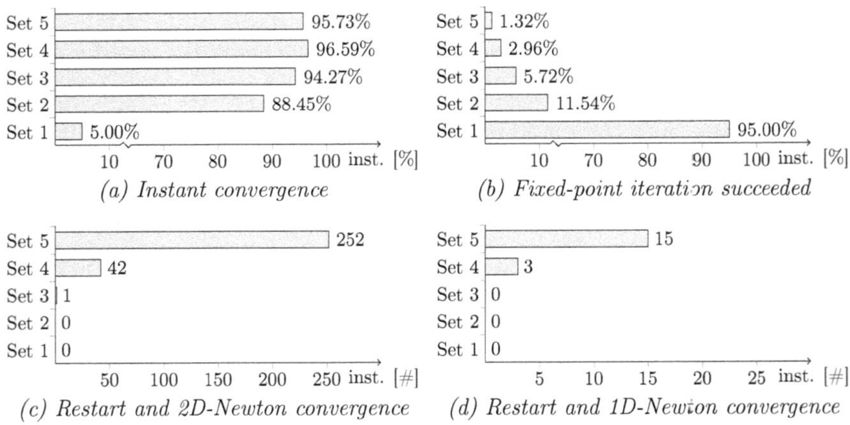

(a) *Instant convergence* (b) *Fixed-point iteration succeeded*

(c) *Restart and 2D-Newton convergence* (d) *Restart and 1D-Newton convergence*

Fig. 5.7: Computational results - CA convergence

Figure 5.7a shows the percentage of instances that converge instantly, i.e. in the first 2D-Newton call, while the portion of instances that converge in the FPI stage is presented in Figure 5.7b. The amount of instances that are solved in the 2D-Newton stage (first 2D-Newton block in Figure 5.5) after FPI has failed, i.e. after "restarting" NM, is shown in Figure 5.7c. Finally, Figure 5.7d presents the number of instances that require a 1D-Newton call to find a degenerate price equilibrium. Note that, additionally, four instances of Set 5 require 2D-Newton calls in the second NM cycle. Furthermore, 24 instances of set 5 do not have a local price equilibrium. The nonexistence of local price equilibria has been manually confirmed for all those instances. It is apparent from these figures that a (combined) increase of the upper bounds on the sensitivity parameters α and β induces the need for more computational effort. Not surprisingly, the number of instances that require a (converging) FPI call decreases over the test sets. This is to some extent because the payoff functions are less likely to be monotonously increasing in the prices of the corresponding players in sets with potentially larger sensitivity parameters (cf. Section 5.1). However, the most striking result to emerge from Figure 5.7 is that CA reliably converges to local price equilibria in case of their existence. Additionally, computational times are rather small as can be seen from Figure 5.8. The figure presents average (Figure 5.8a) and maximum (Figure 5.8b) running times of CA over the instances that have local price equilibria. The figures support the finding that "challenging" instances are characterized by rather large sensitivity parameters.

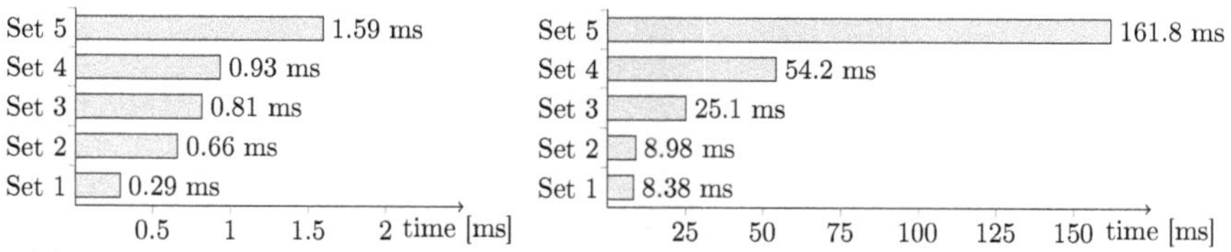

(a) Average computation time, t_{avg} *(b) Maximum computation time, t_{max}*

Fig. 5.8: Computational results - CA running time

Figures 5.9 and 5.10 analyze the effects of increasing sensitivity parameters separately. Each figure relates to six sets (index $i = 1, ..., 6$) of test instances with 1,000 instances per set. p and r are fixed to 5. Furthermore, $\bar{p} = 150$. A specific instance of one set differs from the corresponding instance in another set only in the upper bound (denoted by α_i or β_i) on the range that the "varying" parameter α (in Figure 5.9) or β (in Figure 5.10) is drawn from when constructing the instances randomly (see Table 5.2).
More specifically, Figures 5.9 and 5.10 relate to

- the percentage of instances that converge instantly (Figures 5.9a and 5.10a),

Tab. 5.2: Ranges of sensitivity parameters

i	1	2	3	4	5	6
Varying range sens. param.	$[0.015, 1]$	$[0.015, 2]$	$[0.015, 3]$	$[0.015, 4]$	$[0.015, 10]$	$[0.015, 50]$
Fixed range sens. param.	$[0.015, 2]$	$[0.015, 2]$	$[0.015, 2]$	$[0.015, 2]$	$[0.015, 2]$	$[0.015, 2]$

- the amount of instances that converge in the FPI stage (Figures 5.9b and 5.10b),
- the percentage of instances that converge in the first 2D-Newton phase but not in the first call (Figures 5.9c and 5.10c),
- the amount of instances with degenerate equilibria that are found in the 1D-Newton stage or the alternating 1D-Newton stage, respectively (Figures 5.9d and 5.10d),
- the amount of instances that require calls of NM in the second NM cycle (Figures 5.9e and 5.10e),
- the percentage of instances that do not have a local equilibrium in prices (Figures 5.9f and 5.10f; again, the instances have been manually checked to assure correctness),
- average (Figures 5.9g and 5.10g) and maximum (Figures 5.9h and 5.10h) computational times over the instances with local price equilibria.

Interestingly, while both, an increase of the upper bound on α and the upper bound on β, induce larger average running times of CA (accompanied by a worse convergence behavior), the more severe effect is caused by increasing α's upper bound. Moreover, for $i = 1, .., 4$ we have opposing effects caused by increasing the upper bounds of the customers' sensitivity parameters on the amount of instantly converging instances and the percentage of instances where a local price equilibrium is found in the FPI stage. However, in the case of very large upper bounds on β ($i = 4, 5$), the effect of a decreasing amount ($i = 1, .., 4$) of instantly converging instances or instances converging in the FPI stage, is reversed. Finally, α's effect on the potential nonexistence of local price equilibria is clearly stronger than β's influence.

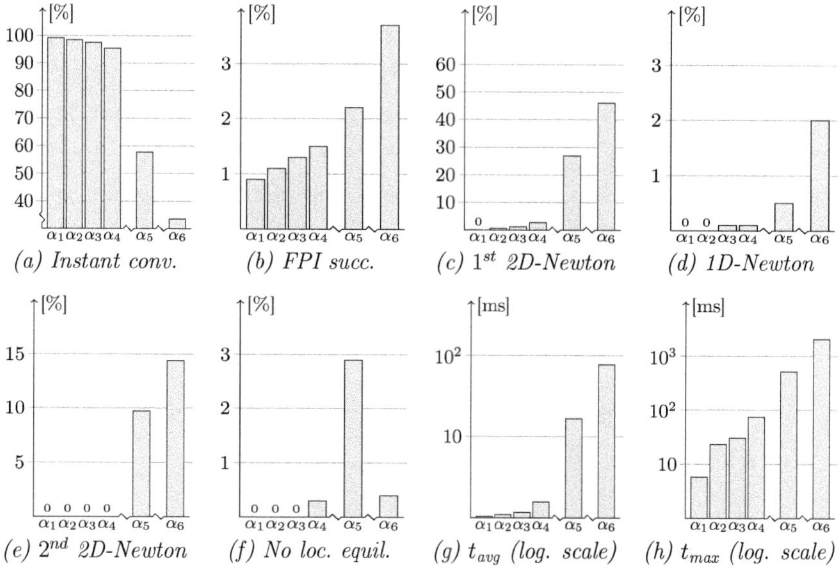

Fig. 5.9: Varying α

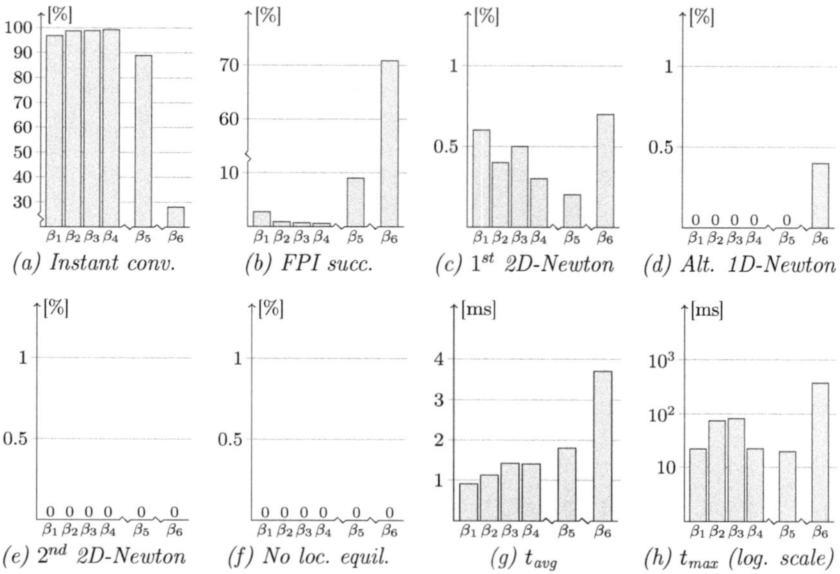

Fig. 5.10: Varying β

Figure 5.11a takes a closer look at α's effect on the existence of local equilibria. Similarly to the prior test instances, we have increased both, the upper and lower bound on α, simultaneously for 1,000 test instances in 16 test sets for three different values of $\bar{p}$ and $\beta \in [0.015, 0.5]$. Again, p and r have been fixed to 5. For each test set, α's lower bound equals the upper bound of the prior test set, with a lower bound of zero in the first set. As already indicated in Figure 5.9f, increasing coefficients of spatial friction at first increase the amount of instances without local price equilibria. If we keep increasing α, however, instances tend to "regain" local equilibria. The latter statement holds for decreasing values of $\bar{p}$ as well.

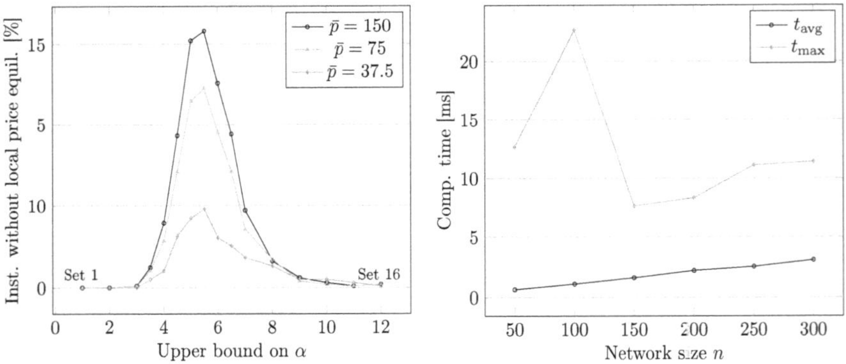

(a) Existence of local price equilibria *(b) Influence of network size on comp. time*

Fig. 5.11: Existence of equilibria and influence of network size

Figure 5.11b presents results on the effect of increasing network size on computational times. The plot is based on 1,000 test instances in six sets with $\alpha \in [0.015, 1.5]$, $\beta \in [0.015, 1.5]$, $\bar{p} = 150$ and $r = p = 5$. Corresponding instances of different sets vary in network size and facility locations. We find a nearly linear increase of average computational times over the network size.

5.3 Location Stage

Benati and Hansen (2002) provide examples, showing that the incorporation of random utility models into competitive location models may actually affect the optimal locations of facilities. It is the aim of this section to show that including an additional pricing game supports the same reasoning.

Let $r = p = 1$, consider the network and the parameters of Figure 5.12 (edge weights correspond to edge lengths) and set $X_1 = \{0\}$ (gray vertex), $\alpha = 1.974$, $\bar{p} = 40$, $c_L = 2.87$, $c_F = 4.1$ and $s = 1$. For all potential follower locations in the network and three different price sensitivity levels, Table 5.3 presents the corresponding unique Nash equilibria in prices[10] and follower's profits. As claimed above, we find a strong effect of β on the optimal follower's choice. While $\beta \approx 0$ induces co-location of the players, larger price sensitivity levels enforce differentiation of locations (in the case of the example instances).

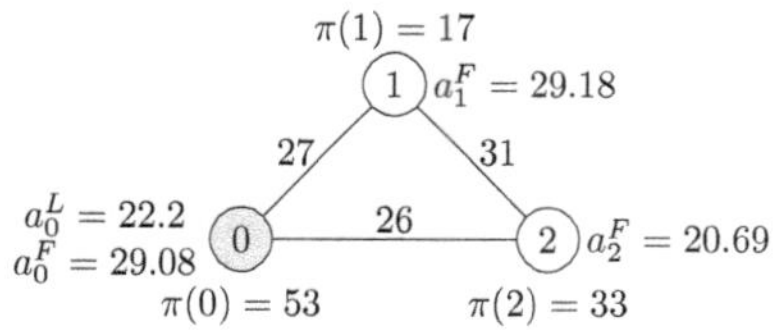

Fig. 5.12: Example network

Tab. 5.3: Optimal follower locations under different price sensitivity levels

β	Follower location	Unique Nash equilibrium in prices, (p_L^N, p_F^N)	Π_F
	0*	(40,40)	1900.52
$1.65 \cdot 10^{-5}$	1	(40,40)	610.227
	2	(40,40)	1184.56
	0	(4.8,13.3)	400.645
0.629	1	(30.8,40)	599.351
	2*	(30.8,28.6)	756.927
	0	(3.4,5.9)	72.531
2.323	1*	(8.5,11.4)	116.205
	2	(8.5,8.0)	114.543

Hence, we may conclude that price competition is worth being considered in competitive location models that utilize random utility models. This, of

10 The uniqueness of the price equilibria has been manually confirmed by plotting the corresponding profit derivatives over the domain of prices.

course, is only true if this is not necessarily accompanied by large computational costs. Therefore, in order to roughly analyze solution times and qualities of related heuristics, we have implemented two straight forward algorithms that apply CA. The development of more sophisticated approaches is left for future research. Especially, such approaches will need to provide adequate strategies to overcome CA's major drawback, i.e. the fact that (in the case of existence of at least one local price equilibrium) it determines one local price equilibrium only. In our analysis, we will refer to a follower's choice of r different locations as a solution to an instance of LPL, if there exists a corresponding local equilibrium in prices.

The *greedy algorithm* proceeds as follows: Initialize $it := 1$ and $Y_0 := \emptyset$ (none of the follower's facilities have been located). Repeat for all potential facility sites $v \in V \setminus Y_{it-1}$: Set $Y_{it} = Y_{it-1} \cup \{v\}$, determine a local equilibrium in prices (if an equilibrium exists) by calling CA, calculate the corresponding follower's profit (if no equilibrium exists, assume the follower's profit to be -1) and reset $Y_{it} = Y_{it-1}$. Eventually, locate j in the candidate location v' yielding the maximal profit, $Y_{it} = Y_{it-1} \cup \{v'\}$. If no local equilibrium in prices exists for any candidate location, choose a random (and feasible) vertex. If all facilities have been located, i.e. $it = r$, and a local equilibrium in prices exists, then stop. Otherwise, if $it < r$ proceed by setting $it = it + 1$ and locating the next facility in the same manner. If $it = r$ and a local equilibrium in prices does not exist, keep generating random sets of follower's locations until a set with an existing local price equilibrium is found. If no such set is found within a time limit t_{max}, then stop.

Additionally, we have implemented a basic *tabu search* heuristic (cf. Glover and Laguna, 1997, for an introduction to tabu search) with the following neighborhood structure: For a set of follower's locations Y_p and for every $i \in Y_p$ and $j \in V \setminus Y_p$, generate $Y_p \setminus \{i\} \cup \{j\}$ (1-interchange moves, denoted by $(\bar{i}, j)$). Evaluate each set of locations by calling CA as in the greedy algorithm. Execute the best non-tabu move. Additionally, apply a simple aspiration criterion, i.e. allow a tabu move if it results in a solution with an objective value that is better than the currently best known follower's profit. If no neighboring set of locations has a local equilibrium in prices, choose a random neighbor. If a move $(\bar{i}, j)$ is executed, record the opposing move $(\bar{j}, i)$ and, additionally, the same move as tabu for tl iterations. Start the procedure by selecting a random set of p different locations. If circling around a local optimum is detected, restart the search process at a randomly generated set of locations (diversification). Terminate the tabu search heuristic after a fixed number of $maxit$ iterations or if the same local optimum has been found $maxloc$ times.

We have randomly generated a total of 400 test instances as described in Section 5.2.4, with $\lambda(uv) \in [1, 50]$ for all $[u, v] \in E$, $\pi(u) \in [0, 100]$ for all $u \in V$, $s = 1$, $\bar{p} = 150$, $c_q \in [1, 10]$, $a_j^q \in [30, 40]$ for all $j \in V$ and $q \in \{L, F\}$. The test instances are grouped with respect to their (randomly drawn) sensitivity parameters, as well as their network sizes and the number of players' facilities; see Tables 5.4 and 5.5. Five instances were generated in each group. The leader's locations were randomly drawn from the vertex set. The tabu search parameters have been fixed to $tl = 25$, $maxit = 100$ and $maxloc = 4$.

Table 5.4 depicts results on the quality of the heuristics' solutions. Solution quality is measured in terms of objective function values (at the corresponding local price equilibria) in relation to the objective function values of the best solutions found by complete enumeration over all potential location settings. However, as before, we apply CA to determine a local price equilibrium for each of these location settings, so that the enumeration algorithm is not guaranteed to find optimal solutions. All cell entries in Table 5.4 are average values over the corresponding group of test instances. Table 5.5 presents the corresponding average computational times.

These basic results indicate that our observations of Section 5.2.4 (reliable convergence behavior of CA within rather small computational times) result in standard heuristics tending to determine high quality solutions within acceptable time, even for "challenging" ranges of the sensitivity parameters. Thus, as claimed above, it seems to be reasonable to include price competition into basic location problems that utilize random utility models. As to be expected, the tabu search heuristic outperforms the simple greedy algorithm in terms of solution quality. Additionally, note that we did not encounter

Tab. 5.4: Average solution quality of location heuristics

			$\beta \in [0.015, 2]$ $\alpha \in [0.015, 2]$		$\beta \in [0.015, 2]$ $\alpha \in [2, 3]$		$\beta \in [0.015, 2]$ $\alpha \in [3, 4]$		$\beta \in [2, 3]$ $\alpha \in [0.015, 2]$		$\beta \in [3, 4]$ $\alpha \in [0.015, 2]$	
n	p	r	greedy	tabu	greedy	tabu	greedy	tabu	greedy	tabu	greedy	tabu
30	3	3	0.92	0.99	0.97	1.01	0.83	1.01	0.92	1.00	0.95	1.00
		4	0.98	1.00	0.93	1.01	0.98	1.02	0.88	1.00	1.00	1.00
		5	0.94	1.00	0.76	1.00	0.90	0.96	0.98	0.99	1.00	1.00
	4	3	0.91	1.00	0.86	1.03	0.97	1.01	0.98	1.02	0.98	0.99
		4	0.94	1.00	0.81	0.96	0.84	1.00	0.95	1.00	0.99	1.00
		5	0.93	0.98	0.93	1.00	0.79	1.00	0.95	1.00	1.00	1.00
50	3	3	0.99	1.00	0.79	0.97	0.86	0.94	0.99	1.00	1.00	1.00
		4	0.99	1.00	0.82	1.00	0.92	1.00	1.00	1.00	0.96	1.00
		5	0.96	1.00	0.76	0.82	0.85	0.98	0.98	1.00	1.00	1.00
	4	3	0.99	1.00	0.95	1.00	0.82	0.93	0.98	1.00	1.00	1.00
		4	0.94	1.00	0.83	1.00	0.87	0.96	0.99	1.00	0.98	1.00
		5	0.97	1.00	0.62	0.78	0.79	0.98	1.00	1.00	0.91	1.00
90	3	3	1.00	1.00	0.98	1.00	0.84	0.94	1.00	1.00	1.00	1.00
		4	0.99	1.00	0.91	1.00	0.93	0.99	1.00	1.00	1.00	1.00
	4	3	1.00	1.00	0.98	1.00	0.92	1.00	0.99	1.00	1.00	1.00
		4	1.00	1.00	0.93	1.00	0.77	0.83	1.00	1.00	1.00	1.00

Tab. 5.5: Average solution time of location heuristics (seconds)

			$\beta \in [0.015, 2]$ $\alpha \in [0.015, 2]$		$\beta \in [0.015, 2]$ $\alpha \in [2, 3]$		$\beta \in [0.015, 2]$ $\alpha \in [3, 4]$		$\beta \in [2, 3]$ $\alpha \in [0.015, 2]$		$\beta \in [3, 4]$ $\alpha \in [0.015, 2]$	
n	p	r	greedy	tabu	greedy	tabu	greedy	tabu	greedy	tabu	greedy	tabu
30	3	3	0.27	14.70	0.22	16.53	0.28	14.20	0.06	5.03	0.03	1.08
		4	0.03	1.07	0.59	48.73	0.45	14.27	0.10	23.81	0.05	3.68
		5	0.03	11.36	0.35	35.82	0.39	34.33	0.08	19.66	0.04	1.63
	4	3	0.05	10.97	0.36	31.35	0.33	9.58	0.54	37.70	0.04	1.91
		4	0.12	10.73	0.35	29.91	0.42	45.60	0.23	51.74	0.19	3.19
		5	0.35	23.04	0.47	40.19	0.36	49.93	0.21	21.97	0.08	5.38
50	3	3	0.06	5.26	0.89	47.63	0.57	27.23	0.06	4.70	0.06	2.73
		4	0.04	2.69	0.75	65.70	1.38	109.03	0.19	5.53	0.10	53.55
		5	0.09	4.63	0.44	170.14	0.94	114.61	0.09	10.41	0.08	2.77
	4	3	0.13	8.06	0.26	14.14	0.77	85.61	0.05	3.84	0.06	3.52
		4	0.11	14.07	0.75	104.09	1.24	127.66	0.07	6.24	0.09	4.31
		5	0.25	15.58	1.64	394.67	0.63	63.16	0.25	14.47	0.22	38.14
90	3	3	0.10	2.80	0.13	10.44	0.35	17.16	0.15	3.20	0.15	8.53
		4	0.13	5.35	0.48	19.54	0.83	185.97	0.21	14.07	0.27	9.04
	4	3	0.13	3.01	0.33	20.72	0.56	66.22	0.18	3.21	0.22	3.59
		4	0.13	3.34	0.15	10.74	1.47	233.67	0.33	5.05	0.25	9.80

test instances without any locational setting with an existing local price equilibrium.

5.4 Conclusion

In this chapter, we have discussed a discrete competitive location problem on networks – the $(r|X_p)$-medianoid problem – with an additional pricing stage (price competition). As in Benati and Hansen (2002), we have assumed customers to be utility maximizers and we have applied the well known multinomial logit approach to model their behavior. Hence, customer behavior has been assumed to be probabilistic.

We have provided insights into the existence of (local) price equilibria and we have proven NP-hardness of the problem. Furthermore, we have shown that including price competition into $(r|X_p)$-medianoid problems under the multinomial logit model does actually have an effect on the optimal follower's locations. Additionally, we have provided a reliable numerical algorithm to quickly determine local equilibria in prices, assuming that the players' locations are given. Finally, based on this numerical method, we have presented first insights into heuristic algorithms to solving the location problem itself.

Future research may focus on several issues. First, adequate strategies to finding global equilibria (Nash equilibria) in prices may be developed. Furthermore, the model may be generalized in multiple ways. For example, one may let the players charge different prices in different locations. Research may also focus on the incorporation of other, more general, random utility models. Other than the multinomial logit model, such models may, for instance, take account of flexible substitution patterns, as their non-consideration is one of the most limiting factors of our model (cf. already Benati and Hansen, 2002).

Chapter 6

Summary and Outlook

This dissertation has investigated the incorporation of proportional choice rules, non-discrete demand, or an additional pricing stage into sequential competitive location problems on networks. We have analyzed the general properties of the resulting models such as, for example, their computational complexity or the existence of finite dominating sets. Furthermore, we have proposed exact and heuristic solution algorithms and we have presented corresponding computational results.

The first two chapters have been concerned with the fundamentals of competitive location theory. Chapter 1 has introduced the basic notation and definitions used throughout the thesis. It has focused on graph theory, game theory and discrete choice theory. Furthermore, a simple classification scheme for competitive location models has been presented. An up-to-date review of literature on sequential competitive location models (including voting location problems) has been provided in Chapter 2. The chapter complements and extends the well known survey by Eiselt and Laporte (1996). Its focus has been on continuous models defined on networks.

Chapter 3 has been devoted to a voting location problem under a proportional choice rule that has previously been introduced by Bauer et al. (1993). We have found that 1-suboptimal points of tree networks need not be vertices and have thus answered a long time open question. However, there may well exist a finite dominating set that future research might explore. Additionally, an interesting question is, if there are simpler network classes, e.g.

chain networks, that posses the vertex optimality property. In the second part of Chapter 3 we have been concerned with implementational issues of an algorithm for determining 1-suboptimal points. Moreover, we have extended the model to include edge demand. This, however, has led to a loss of the discretization properties known from the case of vertex demand only. Furthermore, we have seen that even for the simple class of chain networks with edge demand only, 1-optimal points can in general not be determined by straight forward analytical calculations, so that there remain interesting questions to be answered by future research.

In Chapter 4 we have analyzed the binary $(r|p)$-centroid problem on networks with vertex and edge demand. We have presented bilevel programming formulations for the discrete problem class and we have proposed potential heuristic solution approaches to be examined in future. More significantly, we have proven that the discrete and continuous versions of the binary $(r|p)$-centroid problem are NP-hard on general networks with edge demand only. Moreover, an efficient algorithm has been derived for the discrete problem class with $r = 1$ on tree networks. Future research may focus on several extensions of the model. For example, one may analyze the incorporation of non-uniform demand densities.

Chapter 5 has been concerned with the discrete $(r|X_p)$-medianoid problem under multinomial logit demand. We have extended the literature on this problem by including an additional stage into the game, allowing the players to simultaneously set prices for the homogenous product after having chosen locations sequentially. The problem has been proven to be NP-hard. Furthermore, some general results on the existence of price equilibria have been derived. Hereafter, we have provided a combination of numerical methods to reliably and quickly determine local price equilibria. This method has been included into simple heuristic algorithms for finding solutions to the follower's location problem. We have seen that these heuristics are capable of providing high quality results within acceptable time and that the incorporation of price sensitive customers does actually effect the optimal follower's location choice. However, the follower's location decisions as determined by the heuristics are not necessarily stable, because local price equilibria need not be Nash equilibria. Hence, future research will need to focus on developing strategies to determine or approximate global equilibria in prices. Furthermore, more general random utility models may be incorporated.

Table 6.1 summarizes the main features of the models that have been presented in this dissertation.

Tab. 6.1: Main features of the models presented in this thesis

Chap.	Type of problem	Choice rule	Type of demand
3	(1|1)-centroid (voting location)	proportional	vertex/vertex & edge dem.
4	$(r\|p)$-centroid	binary	vertex & edge dem.
5	$(r\|X_p)$-medianoid, pricing game	proportional (logit)	vertex dem.

Bibliography

Aho, A. V. (2004). Computer representations of graphs. In Gross, J. L. and Yellen, J., editors, *Handbook of Graph Theory*, pages 57–67. CRC Press, Boca Raton.

Aikens, C. H. (1985). Facility location models for distribution planning. *European Journal of Operational Research*, 22(3):263–279.

Aiura, H. (2010). "Signal-jamming" leads to "minimum differentiation" under demand uncertainty. *The Annals of Regional Science*, 45(2):245–271.

Alstrup, S., Gavoille, C., Kaplan, H., and Rauhe, T. (2004). Nearest common ancestors: A survey and a new algorithm for a distributed environment. *Theory of Computing Systems*, 37(3):441–456.

Anderson, S. P., de Palma, A., and Thisse, J.-F. (1992a). *Discrete Choice Theory of Product Differentiation*. MIT Press, Cambridge.

Anderson, S. P., de Palma, A., and Thisse, J.-F. (1992b). Social surplus and profitability under different spatial pricing policies. *Southern Economic Journal*, 58(4):934–949.

Arostegui Jr., M. A., Kadipasaoglu, S. N., and Khumawala, B. M. (2006). An empirical comparison of tabu search, simulated annealing, and genetic algorithms for facilities location problems. *International Journal of Production Economics*, 103(2):742–754.

Avella, P., Benati, S., Cánovas Martinez, L., Dalby, K., Di Girolamo, D., Dimitrijevic, B., Ghiani, G., Giannikos, I., Guttmann, N., Hultberg, T. H., Fliege, J., Marin, A., Munõz Márquez, M., Ndiaye, M. M., Nickel, S.,

Peeters, P., Pérez Brito, D., Policastro, S., Saldanha de Gama, F. A., and Zidda, P. (1998). Some personal views on the current state and the future of locational analysis. *European Journal of Operational Research*, 104(2):269–287.

Bandelt, H.-J. (1985). Networks with Condorcet solutions. *European Journal of Operational Research*, 20(3):314–326.

Bard, J. F. (1998). *Practical Bilevel Optimization – Algorithms and Applications*. Kluwer, Dordrecht.

Bauer, A., Domschke, W., and Pesch, E. (1993). Competitive location on a network. *European Journal of Operational Research*, 66(3):372–391.

Bazaraa, M. S., Sherali, H. D., and Shetty, C. M. (2006). *Nonlinear Programming: Theory and Algorithms*. Wiley, Hoboken, 3rd edition.

Beladi, H., Chakrabarti, A., and Marjit, S. (2010). Sequential spatial competition in vertically related industries with different product varieties. *Economics Letters*, 106(2):112–114.

Benati, S. (2000). NP-hardness of some competitive location models with probabilistic choice rules. *Studies in Locational Analysis*, 14:211–231.

Benati, S. (2003). An improved Branch & Bound method for the uncapacitated competitive location problem. *Annals of Operations Research*, 122(1–4):43–58.

Benati, S. and Hansen, P. (2002). The maximum capture problem with random utilities: Problem formulation and algorithms. *European Journal of Operational Research*, 143(3):518–530.

Benati, S. and Laporte, G. (1994). Tabu search algorithms for the $(r|X_p)$-medianoid and $(r|p)$-centroid problems. *Location Science*, 2(4):193–204.

Berman, O., Drezner, Z., and Krass, D. (2010). Generalized coverage: New developments in covering location models. *Computers & Operations Research*, 37(10):1675–1687.

Berman, O. and Gavious, A. (2007). Location of terror response facilities: A game between state and terrorist. *European Journal of Operational Research*, 177(2):1113–1133.

Birmingham Council (2011). Strategy for special provision (S4SP) guidance. `http://ebriefing.bgfl.org/content/resources/resource.cfm?id=8009&key=&zz=20110130064926142&zs=n`. Last accessed on 08-04-2011.

Bonein, A. and Turolla, S. (2009). Sequential location under one-sided demand uncertainty. *Research in Economics*, 63(3):145–159.

Boyer, M., Laffont, J.-J., Mahenc, P., and Moreaux, M. (1995). Sequential location equilibria under incomplete information. *Economic Theory*, 6(2):323–350.

Boyer, M., Mahenc, P., and Moreaux, M. (2003a). Asymmetric information and product differentiation. *Regional Science and Urban Economics*, 33(1):93–113.

Boyer, M., Mahenc, P., and Moreaux, M. (2003b). Entry preventing locations under incomplete information. *International Journal of Industrial Organization*, 21(6):809–829.

Brandeau, M. L. and Chiu, S. S. (1989). An overview of representative problems in location research. *Management Science*, 35(6):645–674.

Campos Rodríguez, C. M. and Moreno Pérez, J. A. (2000a). Comparison of α-Condorcet points with median and center locations. *TOP*, 8(2):215–234.

Campos Rodríguez, C. M. and Moreno Pérez, J. A. (2000b). The tolerant Condorcet points. *Studies in Locational Analysis*, 14:173–190.

Campos Rodríguez, C. M. and Moreno Pérez, J. A. (2003). Relaxation of the Condorcet and Simpson conditions in voting location. *European Journal of Operational Research*, 145(3):673–683.

Campos Rodríguez, C. M. and Moreno Pérez, J. A. (2008). Multiple voting location problems. *European Journal of Operational Research*, 191(2):437–453.

Choi, S. C., Desarbo, W. S., and Harker, P. T. (1990). Product positioning under price competition. *Management Science*, 36(2):175–199.

Cormen, T. H., Leiserson, C. E., Rivest, R. L., and Stein, C. (2001). *Introduction to Algorithms.* MIT Press, Cambridge, 2nd edition.

Current, J., Daskin, M., and Schilling, D. (2002). Discrete network location models. In Drezner, Z. and Hamacher, H. W., editors, *Facility Location – Applications and Theory*, pages 81–118. Springer, Berlin.

Dasci, A. (2011). Conditional location problems on networks and in the plane. In Eiselt, H. A. and Marianov, V., editors, *Foundations of Location Analysis*, pages 179–206. Springer, New York.

Dasci, A., Eiselt, H. A., and Laporte, G. (2002). On the $(r|X_p)$-medianoid problem on a network with vertex and edge demands. *Annals of Operations Research*, 111(1–4):271–278.

Dasci, A. and Laporte, G. (2005). A continuous model for multistore competitive location. *Operations Research*, 53(2):263–280.

Dasci, A. and Laporte, G. (2007). Erratum to "A continuous model for multistore competitive location". *Operations Research*, 55(6):1188–1190.

Daskin, M. S. (1995). *Network and Discrete Location – Models, Algorithms, and Applications.* Wiley, New York.

Daskin, M. S. (2008). What you should know about location modeling. *Naval Research Logistics*, 55(4):283–294.

D'Aspremont, C., Gabszewicz, J. J., and Thisse, J.-F. (1979). On Hotelling's "Stability in competition". *Econometrica*, 47(5):1145–1150.

Dempe, S. (2002). *Foundations of Bilevel Programming.* Kluwer, Dordrecht.

Dobson, G. and Karmarkar, U. S. (1987). Competitive location on a network. *Operations Research*, 35(4):565–574.

Domschke, W. and Drexl, A. (1985). *Location and Layout Planning – An International Bibliography.* Springer, Berlin.

Domschke, W. and Krispin, G. (1997). Location and layout planning. *OR Spectrum*, 19(3):181–194.

Drezner, T. (1995a). Competitive facility location in the plane. In Drezner, Z., editor, *Facility Location – A Survey of Applications and Methods*, pages 285–300. Springer, New York.

Drezner, Z., editor (1995b). *Facility Location – A Survey of Applications and Methods.* Springer, New York.

Drezner, Z. and Eiselt, H. A. (2002). Consumers in competitive location models. In Drezner, Z. and Hamacher, H. W., editors, *Facility Location – Applications and Theory*, pages 151–178. Springer, Berlin.

Drezner, Z. and Hamacher, H. W., editors (2002). *Facility Location – Applications and Theory.* Springer, Berlin.

Eiselt, H. A. (1993). Competition in locational models. *Studies in Locational Analysis*, 5:129–147.

Eiselt, H. A. and Laporte, G. (1989). Competitive spatial models. *European Journal of Operational Research*, 39(3):231–242.

Eiselt, H. A. and Laporte, G. (1996). Sequential location problems. *European Journal of Operational Research*, 96(2):217–231.

Eiselt, H. A., Laporte, G., and Thisse, J.-F. (1993). Competitive location models: A framework and bibliography. *Transportation Science*, 27(1):44–54.

Eiselt, H. A. and Marianov, V., editors (2011). *Foundations of Location Analysis.* Springer, New York.

Erkut, E. and Neuman, S. (1989). Analytical models for locating undesirable facilities. *European Journal of Operational Research*, 40(3):275–291.

Fleckinger, P. and Lafay, T. (2010). Product flexibility and price competition in Hotelling's duopoly. *Mathematical Social Sciences*, 60(1):61–68.

Francis, R. L. and Goldstein, J. M. (1974). Location theory: A selective bibliography. *Operations Research*, 22(2):400–410.

Friesz, T. L., Miller, T., and Tobin, R. L. (1988). Competitive network facility location models: A survey. *Papers of the Regional Science Association*, 65(1):47–57.

Fudenberg, D. and Tirole, J. (1991). *Game Theory.* MIT Press, Cambridge.

Gabay, D. and Moulin, H. (1980). On the uniqueness and stability of Nash-equilibria in noncooperative games. In Bensoussan, A., Kleindorfer, P., and Tapiero, C. S., editors, *Applied Stochastic Control in Econometrics and Management Science*, pages 271–293. North-Holland, Amsterdam.

Garbe, R. (1995). Tree-width and path-width of comparability graphs of interval orders. In Mayr, E. W., Schmidt, G., and Tinhofer, G., editors, *Graph-Theoretic Concepts in Computer Science – 20th International Workshop, WG'94, Herrsching, Germany, June 1994, Proceedings*, pages 26–37. Springer, Berlin.

García, M. D., Pelegrín, B., and Fernández, P. (2011). Location strategy for a firm under competitive delivered prices. *The Annals of Regional Science*, 47(1):1–23.

García Pérez, M. D., Fernández Hernández, P., and Pelegrín Pelegrín, B. (2004). On price competition in location-price models with spatially separated markets. *TOP*, 12(2):351–374.

García Pérez, M. D., Lafuente Lechuga, M., and Pelegrín Pelegrín, B. (2000). Efficient locations to optimize market share and maximum travel time in a duopoly competition model. *Studies in Locational Analysis*, 14:83–100.

García Pérez, M. D. and Pelegrín Pelegrín, B. (1997). Two-stage competitive location on a tree. *Studies in Locational Analysis*, 11:101–114.

García Pérez, M. D. and Pelegrín Pelegrín, B. (2003). All Stackelberg location equilibria in the Hotelling's duopoly model on a tree with parametric prices. *Annals of Operations Research*, 122(1–4):177–192.

Garey, M. R. and Johnson, D. S. (1979). *Computers and Intractability – A Guide to the Theory of NP-Completeness.* Freeman, New York.

Glover, F. and Laguna, M. (1997). *Tabu Search.* Kluwer, Dordrecht.

Götz, G. (2005). Endogenous sequential entry in a spatial model revisited. *International Journal of Industrial Organization*, 23(3–4):249–261.

Graitson, D. (1982). Spatial competition à la Hotelling: A selective survey. *The Journal of Industrial Economics*, 31(1–2):13–25.

Granot, D., Granot, F., and Raviv, T. (2010). On competitive sequential location in a network with a decreasing demand intensity. *European Journal of Operational Research*, 205(2):301–312.

Gross, J. L. and Yellen, J. (2004). Fundamentals of graph theory. In Gross, J. L. and Yellen, J., editors, *Handbook of Graph Theory*, pages 2–19. CRC Press, Boca Raton.

Hakimi, S. L. (1964). Optimum locations of switching centers and the absolute centers and medians of a graph. *Operations Research*, 12(3):450–459.

Hakimi, S. L. (1983). On locating new facilities in a competitive environment. *European Journal of Operational Research*, 12(1):29–35.

Hakimi, S. L. (1986). p-median theorems for competitive locations. *Annals of Operations Research*, 6(4):77–98.

Hakimi, S. L. (1990). Locations with spatial interactions: Competitive locations and games. In Mirchandani, P. B. and Francis, R. L., editors, *Discrete Location Theory*, pages 439–478. Wiley, New York.

Hale, T. S. and Moberg, C. R. (2003). Location science research: A review. *Annals of Operations Research*, 123(1–4):21–35.

Hamacher, H. W. and Nickel, S. (1998). Classification of location models. *Location Science*, 6(1–4):229–242.

Handler, G. Y. and Mirchandani, P. B. (1979). *Location on Networks – Theory and Algorithms*. MIT Press, Cambridge.

Hansen, P. and Labbé, M. (1988). Algorithms for voting and competitive location on a network. *Transportation Science*, 22(4):278–288.

Hansen, P. and Thisse, J.-F. (1981). Outcomes of voting and planning: Condorcet, Weber and Rawls locations. *Journal of Public Economics*, 16(1):1–15.

Hansen, P., Thisse, J.-F., and Wendell, R. E. (1986). Equivalence of solutions to network location problems. *Mathematics of Operations Research*, 11(4):672–678.

Hansen, P., Thisse, J.-F., and Wendell, R. E. (1990). Equilibrium analysis for voting and competitive location problems. In Mirchandani, P. B. and Francis, R. L., editors, *Discrete Location Theory*, pages 479–501. Wiley, New York.

Harter, J. F. R. (1997). Hotelling's competition with demand location uncertainty. *International Journal of Industrial Organization*, 15(3):327–334.

Hay, D. A. (1976). Sequential entry and entry-deterring strategies in spatial competition. *Oxford Economic Papers*, 28(2):240–257.

Hensher, D. A., Rose, J. M., and Greene, W. H. (2005). *Applied Choice Analysis: A Primer*. Cambridge University Press, Cambridge.

Heywood, J. S. and Ye, G. (2009a). Mixed oligopoly and spatial price discrimination with foreign firms. *Regional Science and Urban Economics*, 39(5):592–601.

Heywood, J. S. and Ye, G. (2009b). Mixed oligopoly, sequential entry, and spatial price discrimination. *Economic Inquiry*, 47(3):589–597.

Hooker, J. N., Garfinkel, R. S., and Chen, C. K. (1991). Finite dominating sets for network location problems. *Operations Research*, 39(1):100–118.

Hotelling, H. (1929). Stability in competition. *The Economic Journal*, 39(153):41–57.

Jaramillo, J. H., Bhadury, J., and Batta, R. (2002). On the use of genetic algorithms to solve location problems. *Computers & Operations Research*, 29(6):761–779.

Kelley, C. T. (2003). *Solving Nonlinear Equations with Newton's Method.* SIAM, Philadelphia.

Kincaid, R. K. (2011). Exploiting structure: Location problems on trees and treelike graphs. In Eiselt, H. A. and Marianov, V., editors, *Foundations of Location Analysis*, pages 315–334. Springer, New York.

Klose, A. and Drexl, A. (2005). Facility location models for distribution system design. *European Journal of Operational Research*, 162(1):4–29.

Kress, D., Li, X., and Pesch, E. (2012). Tabu search for some competitive location problems on networks with vertex and edge demand. Working paper. University of Siegen.

Kress, D. and Pesch, E. (2012a). Competitive location and pricing on networks with random utilities. Working paper. University of Siegen.

Kress, D. and Pesch, E. (2012b). Competitive location under proportional choice: 1-suboptimal points on networks. *Decision Making in Manufacturing and Services*, 6(2):53–64.

Kress, D. and Pesch, E. (2012c). (r, p)-centroid problems on networks with vertex and edge demand. *Computers & Operations Research*, 39(12):2954–2967.

Kress, D. and Pesch, E. (2012d). Sequential competitive location on networks. *European Journal of Operational Research*, 217(3):483–499.

Labbé, M. (1985). Outcomes of voting and planning in single facility location problems. *European Journal of Operational Research*, 20(3):299–313.

Labbé, M. (1990). Location of an obnoxious facility on a network: A voting approach. *Networks*, 20(2):197–207.

Lai, F.-C. (2001). Sequential locations in directional markets. *Regional Science and Urban Economics*, 31(5):535–546.

Lambertini, L. (1997). Unicity of the equilibrium in the unconstrained Hotelling model. *Regional Science and Urban Economics*, 27(6):785–798.

Lambertini, L. (2002). Equilibrium locations in a spatial model with sequential entry in real time. *Regional Science and Urban Economics*, 32(1):47–58.

Loertscher, S. and Muehlheusser, G. (2008). Global and local players in a model of spatial competition. *Economics Letters*, 98(1):100–106.

Love, R. F., Morris, J. G., and Wesolowsky, G. O. (1988). *Facilities Location – Models & Methods*. North-Holland, New York.

Matsumura, T. and Matsushima, N. (2003). Mixed duopoly with product differentiation: Sequential choice of location. *Australian Economic Papers*, 42(1):18–34.

Matsumura, T. and Matsushima, N. (2010). When small firms fight back against large firms in R&D activities. *The B.E. Journal of Economic Analysis and Policy*, 10(1):Article 81.

McFadden, D. (1974). Conditional logit analysis of qualitative choice behavior. In Zarembka, P., editor, *Frontiers in Econometrics*, pages 105–142. Academic Press, New York.

McGinnis, L. F. (1977). A survey of recent results for a class of facilities location problems. *AIIE Transactions*, 9(1):11–18.

Megiddo, N., Zemel, E., and Hakimi, S. L. (1983). The maximum coverage location problem. *SIAM Journal on Algebraic and Discrete Methods*, 4(2):253–261.

Melo, M. T., Nickel, S., and Saldanha-da-Gama, F. (2009). Facility location and supply chain management – a review. *European Journal of Operational Research*, 196(2):401–412.

Mesa, J. A. and Boffey, T. B. (1996). A review of extensive facility location in networks. *European Journal of Operational Research*, 95(3):592–603.

Meza, S. and Tombak, M. (2009). Endogenous location leadership. *International Journal of Industrial Organization*, 27(6):687–707.

Miller, T. C., Friesz, T. L., and Tobin, R. L. (1996). *Equilibrium Facility Location on Networks.* Springer, Berlin.

Mirchandani, P. B. and Francis, R. L., editors (1990). *Discrete Location Theory.* Wiley, New York.

Morrow, W. R. (2008). *A Fixed-Point Approach to Equilibrium Pricing in Differentiated Product Markets.* PhD thesis, The University of Michigan.

Morrow, W. R. and Skerlos, S. J. (2011). Fixed-point approaches to computing Bertrand-Nash equilibrium prices under mixed-logit demand. *Operations Research*, 59(2):328–345.

Nagy, G. and Salhi, S. (2007). Location-routing: Issues, models and methods. *European Journal of Operational Research*, 177(2):649–672.

Nemhauser, G. L. and Wolsey, L. A. (1999). *Integer and Combinatorial Optimization.* Wiley, New York.

Nilssen, T. (1997). Sequential location when transportation costs are asymmetric. *Economics Letters*, 54(2):191–201.

Nilssen, T. and Sørgard, L. (2002). A public firm challenged by entry: duplication or diversity? *Regional Science and Urban Economics*, 32(2):259–274.

Noltemeier, H., Spoerhase, J., and Wirth, H.-C. (2007). Multiple voting location and single voting location on trees. *European Journal of Operational Research*, 181(2):654–667.

Okunuki, K.-I. and Okabe, A. (2002). Solving the Huff-based competitive location model on a network with link-based demand. *Annals of Operations Research*, 111(1–4):239–252.

Osborne, M. J. (1995). Spatial models of political competition under plurality rule: A survey of some explanations of the number of candidates and the positions they take. *Canadian Journal of Economics*, 28(2):261–301.

Owen, S. H. and Daskin, M. S. (1998). Strategic facility location: A review. *European Journal of Operational Research*, 111(3):423–447.

Peeters, P. H. and Plastria, F. (1998). Discretization results for the Huff and Pareto-Huff competitive location models on networks. *TOP*, 6(2):247–260.

Pelegrín, B., Fernández, P., Suárez, R., and García, M. D. (2006). Single facility location on a network under mill and delivered pricing. *IMA Journal of Management Mathematics*, 17(4):373–385.

Pelegrín, B., Suárez-Vega, R., and Cano, S. (2010). Isodistant points in competitive network facility location. *TOP*, doi: 10.1007/s11750-010-0148-6.

Peng, S.-K. and Tabuchi, T. (2007a). Erratum: Spatial competition in variety and number of stores. *Journal of Economics and Management Strategy*, 16(2):537.

Peng, S.-K. and Tabuchi, T. (2007b). Spatial competition in variety and number of stores. *Journal of Economics and Management Strategy*, 16(1):227–250.

Plastria, F. (1995). Continuous location problems. In Drezner, Z., editor, *Facility Location – A Survey of Applications and Methods*, pages 225–262. Springer, New York.

Plastria, F. (2001). Static competitive facility location: An overview of optimisation approaches. *European Journal of Operational Research*, 129(3):461–470.

Plastria, F. (2002). Continuous covering location problems. In Drezner, Z. and Hamacher, H. W., editors, *Facility Location – Applications and Theory*, pages 37–79. Springer, Berlin.

Prescott, E. C. and Visscher, M. (1977). Sequential location among firms with foresight. *The Bell Journal of Economics*, 8(2):378–393.

Press, W. H., Teukolsky, S. A., Vetterling, W. T., and Flannery, B. P. (2002). *Numerical Recipes in C++ – The Art of Scientific Computing*. Cambridge University Press, Cambidge, 2nd edition.

Quarteroni, A., Sacco, R., and Saleri, F. (2000). *Numerical Mathematics*. Springer, New York.

ReVelle, C. (1986). The maximum capture or "sphere of influence" location problem: Hotelling revisited on a network. *Journal of Regional Science*, 26(2):343–358.

ReVelle, C. S. and Eiselt, H. A. (2005). Location analysis: A synthesis and survey. *European Journal of Operational Research*, 165(1):1–19.

ReVelle, C. S., Eiselt, H. A., and Daskin, M. S. (2008). A bibliography for some fundamental problem categories in discrete location science. *European Journal of Operational Research*, 184(3):817–848.

Rhee, B.-D. (2006). First-mover disadvantages with idiosyncratic consumer tastes along unobservable characteristics. *Regional Science and Urban Economics*, 36(1):99–117.

Rhim, H. and Cooper, L. G. (2005). Assessing potential threats to incumbent brands: New product positioning under price competition in a multisegmented market. *International Journal of Research in Marketing*, 22(2):159–182.

Roboredo, M. C. and Pessoa, A. A. (2013). A branch-and-cut algorithm for the discrete $(r|p)$-centroid problem. *European Journal of Operational Research*, 224(1):101–109.

Rothschild, R., Heywood, J. S., and Monaco, K. (2007). Strategic contracts versus multiple plants: Location under sequential entry. *The Manchester School*, 75(2):237–257.

Roy, J. R. and Thill, J.-C. (2004). Spatial interaction modelling. *Papers in Regional Science*, 83(1):339–361.

Şahin, G. and Süral, H. (2007). A review of hierarchical facility location models. *Computers & Operations Research*, 34(8):2310–2331.

Santos-Peñate, D. R., Suárez-Vega, R., and Dorta-González, P. (2007). The leader–follower location model. *Networks and Spatial Economics*, 7(1):45–61.

Serra, D. and ReVelle, C. (1995). Competitive location in discrete space. In Drezner, Z., editor, *Facility Location – A Survey of Applications and Methods*, pages 367–386. Springer, New York.

Serra, D. and ReVelle, C. (1999). Competitive location and pricing on networks. *Geographical Analysis*, 31(1):109–129.

Shiode, S. and Drezner, Z. (2003). A competitive facility location problem on a tree network with stochastic weights. *European Journal of Operational Research*, 149(1):47–52.

Simpson, P. B. (1969). On defining areas of voter choice: Professor Tullock on stable voting. *The Quarterly Journal of Economics*, 83(3):478–490.

Slater, P. J. (1975). Maximin facility location. *Journal of Research of the National Bureau of Standards, Sect. B*, 79B(3–4):107–115.

Smith, H. K., Laporte, G., and Harper, P. R. (2009). Locational analysis: Highlights of growth to maturity. *Journal of the Operational Research Society*, 60, Supplement 1:140–148.

Snyder, L. V. (2006). Facility location under uncertainty: A review. *IIE Transactions*, 38(7):547–564.

Spoerhase, J. and Wirth, H.-C. (2008). Approximating (r,p)-centroid on a path. In *Proceedings of the 7th Cologne-Twente Workshop on Graphs and Combinatorial Optimization, CTW'08*, pages 193–195, Gargnano, Italy.

Spoerhase, J. and Wirth, H.-C. (2009a). An $O(n(\log n)^2/\log\log n)$ algorithm for the single maximum coverage location or the $(1, X_p)$-medianoid problem on trees. *Information Processing Letters*, 109(8):391–394.

Spoerhase, J. and Wirth, H.-C. (2009b). Optimally computing all solutions of Stackelberg with parametric prices and of general monotonous gain functions on a tree. *Journal of Discrete Algorithms*, 7(2):256–266.

Spoerhase, J. and Wirth, H.-C. (2009c). (r,p)-centroid problems on paths and trees. *Theoretical Computer Science*, 410(47–49):5128–5137.

Spoerhase, J. and Wirth, H.-C. (2010). Relaxed voting and competitive location under monotonous gain functions on trees. *Discrete Applied Mathematics*, 158(4):361–373.

Suárez-Vega, R., Dorta-González, P., and Santos-Peñate, D. R. (2007a). The $(r|X_p)$-medianoid model for non-essential services and proportional preferences with attraction functions. *Studies in Locational Analysis*, 16:1–13.

Suárez-Vega, R., Santos-Peñate, D. R., and Dorta-González, P. (2004a). Competitive multifacility location on networks: The $(r|X_p)$-medianoid problem. *Journal of Regional Science*, 44(3):569–588.

Suárez-Vega, R., Santos-Peñate, D. R., and Dorta-González, P. (2004b). Discretization and resolution of the $(r|X_p)$-medianoid problem involving quality criteria. *TOP*, 12(1):111–133.

Suárez-Vega, R., Santos-Peñate, D. R., and Dorta-González, P. (2007b). The follower location problem with attraction thresholds. *Papers in Regional Science*, 86(1):123–137.

Suárez-Vega, R., Santos-Peñate, D. R., Dorta-González, P., and Rodríguez-Díaz, M. (2011). A multi-criteria GIS based procedure to solve a network competitive location problem. *Applied Geography*, 31(1):282–291.

Swamy, M. N. S. and Thulasiraman, K. (1981). *Graphs, Networks, and Algorithms.* Wiley, New York.

Tabuchi, T. and Thisse, J.-F. (1995). Asymmetric equilibria in spatial competition. *International Journal of Industrial Organization*, 13(2):213–227.

Tansel, B. C., Francis, R. L., and Lowe, T. J. (1983a). Location on networks: A survey. Part I: The p-center and p-median problems. *Management Science*, 29(4):482–497.

Tansel, B. C., Francis, R. L., and Lowe, T. J. (1983b). Location on networks: A survey. Part II: Exploiting tree network structure. *Management Science*, 29(4):498–511.

Thomadsen, R. (2005). The effect of ownership structure on prices in geographically differentiated industries. *RAND Journal of Economics*, 36(4):908–929.

Train, K. E. (2009). *Discrete Choice Methods with Simulation.* Cambridge University Press, Cambridge, 2nd edition.

Tyagi, R. K. (1999). Pricing patterns as outcomes of product positions. *Journal of Business*, 72(1):135–157.

Tyagi, R. K. (2000). Sequential product positioning under differential costs. *Management Science*, 46(7):928–940.

von Stackelberg, H. (1934). *Marktform und Gleichgewicht.* Springer, Vienna.

Wendell, R. E. and McKelvey, R. D. (1981). New perspectives in competitive location theory. *European Journal of Operational Research*, 6(2):174–182.

Wendell, R. E. and Thorson, S. J. (1974). Some generalizations of social decisions under majority rule. *Econometrica*, 42(5):893–912.

Yates, A. J. (1997). Hotelling and the New York stock exchange. *Economics Letters*, 56(1):107–110.

Younies, H. and Eiselt, H. A. (2011). Sequential location models. In Eiselt, H. A. and Marianov, V., editors, *Foundations of Location Analysis*, pages 163–178. Springer, New York.

Zhou, D. and Vertinsky, I. (2001). Strategic location decisions in a growing market. *Regional Science and Urban Economics*, 31(5):523–533.

Appendix A

Derivation of the Logit Choice Probabilities

The following derivation of the logit choice probabilities is based on McFadden (1974); Train (2009).

From (1.1) and (1.2) we get

$$pr_{ij} = Prob(\epsilon_{ik} < \epsilon_{ij} + v_{ij} - v_{ik} \; \forall \, k \in J, k \neq j).$$

Hence, with (1.3) and (1.4) and since the stochastic utility components are independent, we have

$$pr_{ij} = \int_{-\infty}^{+\infty} \left(\prod_{k \in J, k \neq j} e^{-e^{-(\epsilon_{ij}+v_{ij}-v_{ik})/\sigma}} \right) \frac{1}{\sigma} e^{-\epsilon_{ij}/\sigma} e^{-e^{-\epsilon_{ij}/\sigma}} \, d\epsilon_{ij}.$$

After substituting $\hat{\epsilon}_{ij} = (1/\sigma)\epsilon_{ij}$ and denoting $\hat{\epsilon}_{ij}$ by z and v_{ij}/σ by $\hat{v}_{ij}$ for all $i \in I$ and $j \in J$ for the sake of brevity, pr_{ij} becomes[1]

$$pr_{ij} = \int_{-\infty}^{+\infty} \left(\prod_{k \in J, k \neq j} e^{-e^{-(z+\hat{v}_{ij}-\hat{v}_{ik})}} \right) e^{-z} e^{-e^{-z}} \, dz,$$

and thus

1 Note that, for readability reasons, we use two identical expressions, exp and e, for the exponential function.

$$pr_{ij} = \int_{-\infty}^{+\infty} \left(\prod_{k \in J} e^{-e^{-(z+\hat{v}_{ij}-\hat{v}_{ik})}} \right) e^{-z}\, dz$$
$$= \int_{-\infty}^{+\infty} \exp\left(-e^{-z} \sum_{k \in J} e^{-(\hat{v}_{ij}-\hat{v}_{ik})} \right) e^{-z}\, dz.$$

Now define $t = e^{-z}$. Then

$$pr_{ij} = \int_{+\infty}^{0} - \exp\left(-t \sum_{k \in J} e^{-(\hat{v}_{ij}-\hat{v}_{ik})} \right) dt$$
$$= \int_{0}^{+\infty} \exp\left(-t \sum_{k \in J} e^{-(\hat{v}_{ij}-\hat{v}_{ik})} \right) dt$$
$$= \left. \frac{\exp\left(-t \sum_{k \in J} e^{-(\hat{v}_{ij}-\hat{v}_{ik})}\right)}{-\sum_{k \in J} e^{-(\hat{v}_{ij}-\hat{v}_{ik})}} \right|_{0}^{+\infty}$$
$$= \frac{e^{v_{ij}/\sigma}}{\sum\limits_{k \in J} e^{v_{ik}/\sigma}}$$

as in (1.5).

www.ingramcontent.com/pod-product-compliance
Ingram Content Group UK Ltd.
Pitfield, Milton Keynes, MK11 3LW, UK
UKHW020121290926
15471UKWH00003B/203

9 783631 627679